FIX YOUR FIREHOUSE

7 strategies that produce a winning team
by
John Lovato, Jr.

Let's Tell Your Story Publishing
London

Copyright

To all those who came before me and took the time to share their knowledge and experiences.

To all those who will come after me and want a way to make their firehouse better than they have had it.

Contents

Foreword

The truth is in the results. The following was given to me by people I have had the privilege of working with. I simply asked, "What do you like about how we do things?"

Oliver C

We keep each other in check. My captain doesn't need to know who is and isn't doing what needs to be done. Firefighters get the grunt work both at the station and in the field.

Engineers make sure our trucks are ready for a call at all times and get us where we need to go safely. We don't always stick to that template. At times, a firefighter makes sure the truck is operationally ready, and an engineer will complete daily chores. We all know how to get the job done. If someone is slacking, they're getting called out. We keep each other in check.

Stress reduction. It's not always there for us, but knowing that a two hour window in the middle of the day is designated to kicking our shoes off is a beautiful thing. Having the busiest zone in the city is rewarding. In contrast, you might lose a night of sleep.

Training. For me it's like going to the gym. I hate it! But when I get done I'm so happy I did it, and I know I gained something from it. My captain knows when to pull the trigger on training. Early in the morning, if calls get heavy and obstacles get thrown at us, he re-evaluates the situation with everyone (We might have to get it done next shift).

Our training isn't bullshit either.

We actually do the things we're going to be asked to do on the job one day. We get together, talk about what we're going to train on, do the training, and then talk about what we liked, didn't like, and what we can do to make it better. Every person in the crew all the way down to the rookie has input. His information doesn't get passed up; if he brings something to the table the crew likes, we're putting it in play.

Leadership. When captain beats you to getting fully bunked out and in the seat for a call, you got yourself a leader. I swear it's a game for me. I have to give 120% to beat my captain to being operationally ready for a call.

There are times I say to myself *I know he knows, this is a bullshit call. Why is he showing off?* But my entire crew and I know that the day we show up and it's not bullshit, he'll be ready to lead us.

Out in the field. The crew knows what to expect from each other. I'll use a car accident as a scenario for out in the field. When the tone drops, we all analyze dispatch in our own ways, determining the severity of the call our rescue might join. On the way to the call, captain is verbally tossing around information about location for the engineer and hazards and injuries for the firefighters.

Pulling up, our engineer is going to make sure he positions the engine so it protects his crew at work. Hopping out, captain is doing a 360 while firefighters are heading straight to the patients to see what injuries need to be taken care of. Engineers are constantly gathering information for their report writing and seeing if firefighters need any equipment from the apparatus.

Once EMS is on the scene and patients are transferred to their trucks, firefighters know to clean up hazards and, if possible, get vehicles off the road way. To drivers breaking their necks as they pass the scene it looks chaotic, but my entire crew is on the same page and getting the job done.

Rules of engagement. The simplest thing every captain can possibly do is tell his crew what he expects from them. This is a one-time thing. When a new crew member arrives to the station, the intercom goes off at some point throughout the day, and that new guy is going to the captain 's office. *Here's what I expect out of you.* Communication is key.

10 for me, 10 for you. There's only one officer on our department who has sat down with each crew member with this question. What are 10 things you will bring to the table every day and what are 10 things you want me to come with? Telling him what I was going to bring to the table was easy, but I actually had to go home and think about what I expected from my captain every shift. And then present it to him. Once again, communication is key.

Daily assignments. We all know what to expect from each other. At the beginning of each shift we have this corny ass dry erase board by the recliners that states your position on the apparatus and what tools you're handling on a fire. It's corny as hell, it's so damn simple, but I love it.

If I'm told at the beginning of the day that I'm in charge of grabbing the irons, I'm not going to forget. When the tone drops, the firefighter doesn't run to the dry erase board to make sure he's the nozzle guy. But it's a constant reminder of *don't get off task.*

When Position 3 gets to the door with the nozzle, he knows Position 4 is behind him with the irons, and they both know rescue 13 crew is coming with a hook and assisting in search and rescue. I'll never forget the time we pulled up to a storage unit with light smoke showing and I was driving Rescue 13.

I neglected to grab the hook even when my firefighter reminded me to grab it. I didn't need the tool on the job, but that's not the point. It's easier to show up and bring to the door your assigned tools; if you don't need it, set it down outside the door. If I had needed the hook, I would've run back to the truck like an asshole wasting time.

Family. It's important to understand that not every shift on every crew member brings 100%. I feel our crew understands this concept and each individual is aware when someone's tank is running on empty. We all know you're getting called out when this occurs, but we also know the person exposing your lack of motivation is going to hump a little bit harder until you get back in the game.

We spend a third of our lives together. We all bring stories from home whether good or bad, and everyone is there to listen. I know the ones in my crew I can count on with advice, or the ones I can count on to carry a bit more of the load until I get things in order. I know this is the way I look at it, and my goal is for every crew member down the line (I don't care if your first shift was last week) to understand they don't have to bottle anything up for 24 hours.

The vulnerability in every one of our crew members is what makes us family.

Delegating responsibilities. I'm a huge advocate of delegating responsibilities. I feel it promotes confidence and keeps complacency away from our crew. One of the first responsibilities I really wanted to take over was report writing. I wasn't trying to change anyone's view on who should be doing reports, but I knew I wanted to take care of my driver who was running 15 calls a day.

When we implemented image trend throughout the department, it became even easier to get reports done before returning to the station. While my driver is getting me back home safely, I'm completing his report as we pull up on the apron. This is an idea that has spread throughout our crew. Now getting ran into the ground, as an engineer on R13 is a little easier.

Morning checklists have been delegated to the rookies. Yeah, engineers need to oversee things are where they need to be and medics need to bless the ALS bags. But our rookies checking our trucks from top to bottom and front to back give them that confidence of knowing the compartments when shit hits the fan.

Conducting a pre-plan. I had no idea what the hell I was doing until my captain slapped me with the responsibility. Giving the firefighter the job of a pre-plan is brilliant! Making contact with business owners in the community, setting up appointments to do a walk through, and mapping out target hazards, alarm panels, fire hydrants, and means of access to an emergency are only bettering us as a crew.

Work hard, play hard. It's no secret, we're the busiest station in the department. We run the most calls every year. We're the ones that train the hardest. But as I always say, we're the ones with all the good stories. With the tones raining down on us all day the crew definitely knows how to play hard. If one of us is doing it we're all doing it. Once we complete tasks for the day it's either basketball, ping-pong, movies on the box, or fire training videos. We play hard, but it always reverts back to training.

Brad D

First, let me introduce myself. My name is Bradley. I am a probationary firefighter. I have had the fortunate opportunity to have Captain Lovato as my captain for a few months. I also was lucky enough to have him as an instructor in the fire academy for a couple live fires and other fire ground training.

When I first got hired, I had my orientation with human resources and was sent right to Station 1 in the heart of downtown; this is where our administration is located. When I showed up and reported to the captain, he was shocked to see me. He hadn't been given any notice that I had been starting today; to his knowledge I was supposed to start the following week.

I really didn't get any orientation with the department or training on anything. They just assigned me structural fire gear and a face piece and informed me that I would be running calls with them today on the engine. I was told to do three eight-hour days to make up the twenty-four hours. I then got assigned to "A" shift. I wasn't assigned a particular station the first few weeks, so I had an opportunity to see how other stations went about their morning routine.

When I arrived to work, I did what any new firefighter would do. I showed up more than 30 minutes early and helped get ready for the day, from making coffee to performing truck checks.

My first shift at Station 3 I did what I did at any of the other new stations I went to. All the stations had a morning meeting, but the one at this station was different. Captain made sure everyone was on the same game plan for if and when we got a call. This was a refreshing experience, especially for a new guy at another different station. I thought that was very unique. Since I was the new guy when I was at a new or different station I wanted to know if something were to happen how we would go about handling it. The disappointing response that I received was often, we will figure it out. When we get there, everything is different.

Although I have received training, every department's procedures were different. I never wanted to be the new guy asking what to do when I got to a scene where everyone was looking at us to act. I found that Station 3 always had answers for my questions, and if not, we would train to answer them.

I enjoy that this station trains for a variety of situations and as a probationary firefighter I feel it is helpful when we get a call because of our training and working together. Having this training and communication helps everyone be calm and react fast when we need to.

I was asked one time when I was on shift to go Station 6 to cover the first 12 hours. The way the calls and dinner worked out, I didn't make it back to Station 3 till 9 p.m. After preparing my gear, par tags, and bedding it was fairly late. I had expressed my interest to captain before I left for Station 6 that I wanted to fix the forcible entry door. I had made a comment when I got back that it was way too late to be trying to do that now. I was surprised when captain was still willing to go out and work on techniques with me. We worked on entries by yourself and with another firefighters. Not only do I find that this station is unique because of captain's communication and training but also with the willingness to make each other strive for the best we can be.

This example is only one of the many ways captain strives to make his team better. Captain Lovato also wants to make sure that he gets the best out of his firefighters, not just training with us, but being able to see if we can make decisions on our own. Once, during training he told us to pull a dry line to a point of entry to see if we pull the correct hose. Captain also did an exercise with us called "10 and 10" in which he gave us 10 expectations and we had to answer how we are going to meet them. The answers to these expectations were goals in how we would meet these expectations.

Just in the few months that I have been at the station, I can see that Captain Lovato strives to better himself as a captain , a leader, and as a firefighter. There is not one 24 hours shift in which we don't get high quality training. Captain is a constant reminder to do our job, treat people right, give an all-out effort with an all in attitude. Because I am not here for me, I am here for we and we are here for them.

Introduction

About me

Hi, I'm John. I grew up in the fire service. I started my career on the same fire department that my grandfather and father worked for.

My current practices are a hodge-podge of what I have seen get the best results over the years. These practices were learned from former company officers I have had and other mentors in the fire service that I have had the privilege to learn from.

About this book

This book contains seven strategies that I use successfully in my firehouse. These are not theories. This is experience-driven advice that can be found in other firehouses with a similar culture of wanting to be the best.

Why I wrote this book

We do a lot of "on the job" learning. We mirror those who came before us since they were our influence. Well, what if you know there is a different way but have not seen it in action yet? You have to go find it. Since 2010, I have actively sought out fire service leaders around our country, hoping to gain a few nuggets of knowledge from each encounter.

This knowledge has led to my daily practices around my firehouse. In 2015, I put out an online survey asking a variety of questions to a Facebook group of firefighters. After reviewing the feedback, I was surprised at what I read. None of the issues that were being talked about existed in my firehouse. That's when I realized I needed to share what we were doing.

What this book covers

I don't work in a fire station, I work in a firehouse. A firehouse has a family environment full of confident firefighters skilled in their craft. This book contains the meat and potatoes of achieving this same environment for you.

Strategy 1: If not you, then who?

If you don't step up and take on the responsibility then who will? You can decide to sit back on the sidelines. However, if you do, don't complain about the performance of the team.

Strategy 2: Get everyone on the same page

A team in the firehouse is a team on the fire ground. This is accomplished by everyone knowing what the plays are. We must know what we must do and what the person next to us is going to do.

Strategy 3: Firehouse family

There are similarities to being a family at home and a family in your firehouse.

Strategy 4: Your time is valuable

It's not about having time, it's about *making* time. You need to set priorities; resources are not limitless.

Strategy 5: Operating at Peak Performance

Got kids who nap? How do kids act when they miss a nap or when it's close to bed time? They probably start getting crabby and irritable. They are exhausted and need rest. We are the same way. When basic needs like hunger and sleep are not met, we are more irritable and less on our game.

Strategy 6: The secret to increased morale

Have you ever seen a happier crew than one who just caught a job? Doing our job makes us happy. We feel value when we perform the skills we specialized in. Companies who catch fire regularly start seeing a decline in morale after a few shifts without a fire. When was your last fire? A week, month, year ago? How can we keep people motivated and committed to the team with long spans between catching a fire?

Strategy 7: Leave the fire service better than you found it

Our time here at the department is leased. It has an expiration date. The department was here long before us and will be here long after us. You are either improving it or weakening it. There is no in-between.

How to use this book

This book is a collaboration of what works. My goal was to have as little fluff as possible and pack it full of great takeaways. Use some of it or all of it.

- write in it
- highlight key points

- bend the pages

A beat-up looking book means it's been referenced a lot and considered valuable.

Here's the deal: learning best practices is great. The real change comes when you implement what you have learned. Too often we just get caught up with learning and never take action on what we have discovered.

Take action as quick as possible. Knowledge is useless unless acted upon.

Download the resources

Head to the site below for access to the resources from the book:
http://www.brotherhoodcoaching.com/fix-your-firehouse/

Keep in touch

If you want to contact me, or join my Facebook group, you'll find details at:
http://www.brotherhoodcoaching.com.

The best leaders are not driven by ego or personal agendas. They are simply focused on the mission and how best to accomplish it.
Jocko Willink

Strategy 1
If not you, then who

What does this mean to you?

If not you, then who? You may have already heard this phrase sometime in your fire service life specifically from people of higher ranking than you. This was a common phrase repeated by one of my battalion chiefs.

The key point is, if you don't step up and take on the responsibility then who will? You can decide to sit back on the sideline, but don't complain about the performance then.

Think of your favorite sporting event. The batter struck out. The kicker missed the field goal. The receiver fumbled the ball.

Whatever the scenario, this is when you sat there on your ass and said, "Well, if that was me, I would've scored."

That's easy to say when you are sitting there on the sidelines of life and not involved in the game. This holds true to watching sporting events and in your everyday life. How many times have you not been involved in putting on a training, running a scene, running a firehouse, or making administrative decisions yet sat there and criticized those who were? So how about we all get involved and make things better?

This is the whole idea behind the phrase *be the role model you looked up to*. This is my personal mission statement, which is also the slogan for Brotherhood Coaching.

The rest of this chapter will cover being the role model you looked up to, what holds you back, common mistakes people make, how to start maximizing your potential, and how to start inspiring the best in others.

Develop a mindset of ownership

The definition of insanity is doing the same thing over and over and expecting different results. This isn't something new that you have never heard of before. How many times in your fire department have you kept doing the same thing yet complain about change? The culture of your station or your department will not change until you do something different. You want to lose weight, work out harder and eat better. You want people to perform better, train more.

If what you're doing isn't working, make the changes.

This could be changes that you see above your rank, or they can be changes you want to see below your rank. These are all areas you can influence by having ownership mentality.

Ownership mentality will enable you to take responsibility for those under your watch and have you take an influential approach for those above you. If you're not happy with the results you get from your supervisors, change the way you present your argument.

Present your case in a way to make them want to do what you're suggesting. You might have to actually do that work, but too many times people bring problems to people above them and not wanting to put the work in themselves.

If you want people under your watch to change what they're doing, be clear on what you expect. If they're not accomplishing what you want, take a second and start thinking maybe it's the way you're communicating it. This applies to your home life. If you have kids, you've probably realized they all respond differently. So do adults. Learn how to get through to them.

We get out of life what we put into it. The results you have right now are the results of the hard work you've done over the years. If it were easy, somebody else would've already have done it. This is the phrase I've heard many times over my career as a firefighter.

Here's an example of influencing ranks above you.

I've always liked the fact of celebrating accomplishments. Too many times a promotion goes unnoticed or not celebrated the correct way.

Let me ask you this

- how were you hired or how were you promoted
- were you hired with a phone call and then tossed on shift
- were you promoted with an email and tossed a badge through inter-office mail

How did that make you feel?

One of my good buddies received his captain's badge through the inter-office mail. His promotion was a big accomplishment that he earned. He saw little thought put into his accomplishment by his superiors by the way he received his badge.

What kind of message is this sending your people?

I always liked how a lot of departments have done hiring ceremonies or promotional ceremonies to celebrate the accomplishments of their people. My department didn't have a formal ceremony that was held regularly.

So, I set out to start one.

I requested to meet with all of our chiefs to make a presentation to start a formal ceremony. In the presentation there were examples of how recipients felt by showing photos of fathers and sons passing on tradition. The celebration had a family atmosphere.

Every chief saw the value in holding a ceremony for the members. Everyone started brainstorming on how to go about it. Like most fire departments, my chiefs wear many "hats".

The one hitch was that everyone wanted to know was who would be heading up this project. I felt like a lot of the plans on how to hold the ceremony were complicated. The more complicated something is, the less likely it will happen. I soon realized that since I brought forward the idea, then I should be the one to coordinate this project. Everyone seemed to like this idea, and I learned a valuable lesson.

When you bring solutions to problems, then you better be prepared to do the work.

For a few years now, I have been coordinating our hiring/promotional ceremonies. It was a little nerve-racking putting together our first one. It came together well, though, with the help of our administrative staff. During the first one, I was standing in the back watching everyone receive their badges. It was a great experience to watch the newly hired and promoted, and their families all gathered together to celebrate their loved ones' achievements. Imagine being able to recreate the feeling you had as a kid on your birthday or Christmas. That's the point of celebrations, to make people feel loved and valued. It takes work to put any party on, but it's always worth it in the end.

Next, I'll talk about owning the changes of those you lead.

Teaching at the fire academy and doing daily trainings with my crew has taught me a lot on teaching for results. I learned when I was out at the fire academy that trainees have very little knowledge of the fire service. Some perform certain tasks poorly and some perform it well. I never wanted to be one of those instructors that would just yell at kids who didn't perform well.

A lot of times the underlying issue is just a misunderstanding.

I remember attending a fire academy and being yelled at for shutting down a hose line. The instructor talked to me like I was a dumbass until he found out that he was the one who hadn't let me know that I was not supposed to shut down the line.

This happened over fourteen years ago, and I remember it quite well. So my goal was not to be the yeller.

Get in the habit of *asking* and not assuming.

If someone you are leading or instructing does not perform a skill properly, ask why they performed it the way they did. You'll often find that there was a misunderstanding or a lack of knowledge on why to do the task a certain way.

Own the mistake since you were not clear enough on the task.

One way to solve this is to have people repeat back the directions you gave them. This confirms an understanding of the task. A clear understanding of *why* the tasks needs to be performed is also helpful.

I'll explain.

I assign the firefighter on our rescue, the hook. The majority of our fires occur in older Florida homes with tongue and groove ceilings. It is rare for a fire to extend into the attic.

Most firefighters in my department only see the hook used in overhaul situations. On a recent fire, the hook was not brought and was left in the rescue.

When I asked the firefighter why he chose not to bring the assigned tool, he replied that he didn't see it being needed. I had two choices to respond:

- I could get upset for him not following orders
- I could ask why

So I asked.

He had never experienced an attic fire, so he had no understanding of the point of grabbing the tool. We discussed the benefits of having a hook readily available.

We were seeing new construction homes pop up in the middle of older neighborhoods. I told him story of a fire I had in which we had to get into the attic quick and having a hook available was crucial. Now we were on the same page.

Your experiences are not always their experiences.

There will be times when a little extra time is needed for explanation. This will go a long way with a person. The 'cause I said so' should be kept for children under the age of five.

Check your ego at the door

Egos eat brains...

Egos eat brains...

and one more time to fully get my message across

Egos eat brains!

Remember this simple phrase consisting of three words. We all have egos. Accept it and work to not let your ego screw things up for you.

An ego is seen in a new firefighter who has been on a couple months longer than another firefighter. They see themselves as being "senior" to this person. Really? Tell me more how much more "seasoned" you are. How about the officer who always points to his collar when his decision is being questioned.

The ego messes with chiefs when they close off their thinking to what they want as opposed to the firefighters they are supporting. Do some internet searches for fire departments firing chiefs, officers, and firefighters. The root cause will typically be the person's ego.

When we put our ego or pride aside, we open up our minds.

We all grew up differently and have had different experiences in our lives. Everyone brings a perspective to the table. Just because someone is younger or has less time on does not mean they have nothing to offer. You would be surprised on what you can learn from someone when you check your ego at the door.

I recently worked with two probies. One used to be a lifeguard, and the other worked on fire alarm systems. We had the lifeguard teach us some water rescue techniques, and the fire alarm tech shared his knowledge with us on fire alarm activations. I learned more about fire alarms from him in a few months then I have in my career.

Imagine if these probies were told to not *talk* and to just *listen*.

Would they be reluctant to share their past experiences with us?

Absolutely.

Think about times you prejudged a person even though you don't know them yet. You "*heard*" from another station that they "*just don't get it*". Odds are you are going to treat them like they don't get it.

People don't become better when we treat them poorly.

I had two different experiences when I first got on the job where I was prejudged.

"HEY, SUPERMAN"

I was working on two different fire departments that had two stations each. Let's call one Fire Department A and the other B. Both positions were part time positions so combined; I was able to make full time pay.

I was young and quiet. You remember just starting out, not knowing how to take everyone, and just feeling everything out in the beginning. I only worked for Fire Department A and B for a couple weeks.

I had just finished a transport to the hospital while working for Department B. A firefighter for Department A was also at the hospital. I only knew this because of his shirt. I had not yet met this person.

While in the medic room finishing my run report, this other firefighter shouted, "Hey, Superman!". Now, I have been called many things before but never Superman. So I assumed someone was by the door or something.

Well, I was wrong.

The firefighter shouted again, "Hey, Superman, I hear you're the best and know everything!" Oh, how I wish I could've told you I responded with a witty comment like "Your mom seems to agree", but I was much more reserved and quiet back then.

I replied that I did not know what he was taking about. He didn't seem to like this response and was getting worked up like a monkey about to throw poop, so I left.

Why did someone who not know me act like this? I was shocked. If I was doing something wrong, why didn't anyone tell me?

Well, when you put a bunch of alpha males who are typically poor at communicating into a group, you get this masculine form of a soap opera. I've always liked the term *As the Hydrant Turns* to describe this phenomena.

Alright, so that was my first experience with no one being direct and passive aggressiveness in the fire service.

The next experience I had was at Fire Department B.

"SHIFT EXCHANGE"

I did a shift exchange with a guy on another shift and station. I was working with a lieutenant who I never worked with before.

At the end of the day the lieutenant took me aside and told me that he had heard some "*negative things*" about me. He stated that he wanted to see for himself and that nothing he had heard was true. I respected his directness and honesty.

Now both experiences involved people creating their own version of things about someone without really understanding the whole story.

Has this every happened to you? Have you ever done this to someone?

Yep. It's human nature.

When something happens, we interpret our own story on why it happened. We also like to find ways we are better than others. Just be aware of this habit. You can act like the asshat in example one and stick with pre-judging someone, or be like the lieutenant in example two and ASK!

This rolls into the good practice of *listening and learning from everyone on your crew*. People want to be engaged, they want to be involved.

If they aren't involved it's due to poor leadership in their past. Leadership that let them down, shot down every idea they had, asked for their input then did their own thing.

We are shaped by the experiences we have had. Show people you listen and value their input, and they will eventually start giving more.

A real easy way to start is to try out suggestions they offer. Even if you know they won't work. Let them give it a shot. Maybe it didn't work before because of timing or if they tweak the way they do it this time, it will be successful.

"CLEVELAND LOAD?"

We were putting together an extended hose load (courtyard load). The load was consisting of 300' of 2.5" hose with 100' of 1 3/4".

Two of the shifts were testing out different ways to make the most out of the load.

One of my guys wanted to try out the Cleveland load. He had been researching it and was eager to test it out.

My department had tested this load out years prior and had discounted it.

I had a choice. I could tell him that no one would be on board and to not bother or say fuck it, let's do it.

So, we put the Cleveland load together and tried it out. It worked well and we liked it.

We did, however, see the limitations it had compared to a shouldered bundle load that we were also working with.

After trying both loads out, we came to our own conclusion that we would stick with the shouldered bundle load.

The end result was that we did not use the suggested load. However, the firefighter was able to give it a shot and see for himself the limitations instead of just being told them. This firefighter continues to make suggestions and was never made to feel that his input was not valued.

Everyone has something to offer.

I don't care how many bugles you have, it does not mean you are better or are more valuable than a person with none.

Limiting beliefs are invisible handcuffs

Limiting beliefs hold us hostage in our own lives.

What if I told you that whatever you tell yourself will happen?

If you tell yourself you can't do something, you won't be able to do it. If you tell yourself you can do something, you will be able to do it.

Telling yourself you will be unsuccessful how limiting beliefs start to come up to the surface.

Look around your personal life and your professional life right now. How many people in both these worlds always make excuses why they are not succeeding at their goals?

We tend to talk ourselves out of things and make excuses why we couldn't achieve them. Examples of this are promotional exams.

There are times when senior members of a department are intimidated by the study habits of junior members. So instead of studying for the exam they occupy themselves with busy activities around the station to make excuses. The crazy thing is the younger generation is not smarter – they are just hungrier.

It all comes down to the willingness to put in the work. As far as promotional exams go, if you're not putting in 40 hours a week of studying then you are not going to be number one.

My first captain's exam I took I scored number five. The top three got promoted. I thought I put as much work as possible. My written score was in the low 70s, and my practical assessment was in the high 80s. Before the next captain's exam came around, I decided this would be the test I get promoted on. I would do what I needed to do.

So I started looking into testing companies and I started talking to people who had scored number one on their test. The common factor was they all put in 40 hours of studying each week. Their scores had more to do with the work they put in than their actual intelligence.

So that's what I did. I was not able to study much on-duty because of the call volume. The majority of my studying was off-duty. I put in 40 hours a week of studying. When the results came out, I scored number one. My written score was in the low 90s and my practical was in the high 80s.

Too often people take one promotional exam, don't do well, and never take another one.

Success comes from failures.

We all fail.

Don't let the failures define you. Success comes from keeping forward momentum. You might get slowed down, but keep moving. You only really fail when you stop moving.

Be your own hero

Whatever it is that you want to see accomplished, you have to be the one who does it. If you want

- a station patch for your crew design it
- more training, attend more trainings

Too often we think the person above us or multiple people above us are the ones to solve all the problems.

The true problem solvers are those on the floor. You don't need permission to be good at your job. Chief officers above us are unaware since they do not get exposed to what we are experiencing on the floor anymore.

Think of it this way – if you're a firefighter you can send an issue up the chain of command. As the issue climbs the hierarchy it starts getting separated from people actually involved with the problem.

When it goes to your captain, that's one person removed. It goes your battalion chief, it's two people removed. Every rank it goes higher makes it even further away from the people actually affected.

It's best for those being affected to handle it themselves. Simply, there's less of a care factor as things climb the chain of command. The more involved you are in the issue, the more emotion you have invested in it.

No one is coming to save you

Your life is 100% your responsibility. If you're thinking that you'll wait for somebody else to make all the problems that you're having go away, then you're going to be waiting a long fucking time.

- the only person holding you back is yourself
- if you want to see improvement, you have to be the one to do it

Rid yourself of the rules that stop you from doing things. How many great ideas have you had that you talk yourself out of?

What are the reasons for talking yourself out of these things?

- I don't have the authority
- I need to be a captain or chief to pull this off

These are all made up rules in your head. We do this so often that we talk ourselves out of working out, eating right, learning more, and making a positive difference.

Forget what people might say about you

Do you hold back from something because you are afraid of what people might say about you?

This is got to be the dumbest reason not to do something.

Stop giving a fuck what other people think about you. If what you're doing is the right thing to do, then do it.

You will never make everyone happy. You can give six people $100. Two will be very appreciative. Two will complain it was not $120. The other two will talk shit about you because they made up a story in their head that you think you're better than them. You're damned if you do and damned if you don't.

Those who work hard and have a good family life will be appreciative of things that are done to make a positive impact. Those who are struggling tend to have a lot of issues going on, and always seem to be the ones to talk shit about you so they can feel better about themselves.

If what you're doing makes things better, then why do they care what you're doing? Odds are they would like to do what you are doing but don't have the courage to do it.

It's very similar to the phrase *haters gonna hate*.

A lot of it truly comes down to just not understanding. To make the changes you want to see you have to let go of what other people think about you.

To avoid criticism say nothing,
do nothing, be nothing.
Aristotle

And that's exactly what a lot of people do.

Accept and get over the fact that people will talk about you. Then you can just move on because what they think of you has no bearing on your feelings. You choose whether or not to react to what happens to you.

It's ultimately our choice to react negatively or positively to what happens to us.

"AMBASSADORS"

I recently had quite a few crew members removed from my station and assigned elsewhere. I like creating a family environment in my firehouse.

Having these people removed from my crew bothered me a lot. I've invested a lot of time in them, and we've grown closer over the time spent.

I spoke about this with a trusted mentor of mine. He's dealt with similar situations throughout his career. He was able to put it in a different perspective that made me look at the situation in a positive light. He told me that I should look at my crew members as ambassadors to what we've been doing. Now they can go represent what we've been doing at other stations.

By spreading out the good we can influence more.

This makes a lot of sense to me. To see the culture I developed spread, I need to have it expand outside the walls of my firehouse.

By changing my perspective, I CHOSE to respond to the change in a positive manner.

Evil only wins when good people do nothing.

If the good people on your fire department or in your firehouse are not spreading what they're doing then the lazies will win.

I worked too damn hard to let the lazy win... and I'm too stubborn.

To break off the handcuffs that you put on yourself abolish the limiting beliefs that you've made up in your head. Start acting like a role model that you look up to when you got on this job.

You are a dad, coach, friend and boss

You have to wear multiple hats. You don't have to be the officer to fill these positions. However, someone has to in order to make your firehouse a success. The more people feel taken care of and safe, the better they will perform.

Be a dad

An officer who takes ownership starts off by filling the dad role in the station.

As a dad, you take care of the crew – protect and prepare people in the short term. Protect your people by

- giving them relevant training
- educating them as much as possible on their job
- sticking up for them

You will be the one who notices their deficiencies first. Recognize these and help them become better.

As a dad, you also prepare your people. Give them more responsibilities around the firehouse and set them up for success. Delegate tasks with value. A person will feel accomplished after they deliver a monthly training or complete a preplan. This will have a greater impact than something like ensuring the toilets are clean. Gauge each person's ability and give them a task they can accomplish. These little wins will pave the way for success down the road. Prepare them for the high risk low frequency calls. You will not always be around, and the more prepared they are the better off they will be.

Be a coach

Develop them for the long term. Everyone needs a coach. There are so many distractions in life that it is easy to get off task. A coach will help their crew develop goals and attain them. You get better results by encouraging people than holding them back. Be there for your people.

Be a friend

Hang out and have fun, be on their level. It's ok to be friends with your crew. Whoever said you couldn't sucked and made up a reason not to have a strong relationship with their crew.

We are social creatures and happiest when we are with like-minded people. Go fishing, go out to lunch, grab some drinks... whatever you have in common, go do.

Be a boss

Take the lead and make decisions. There are times you have to be a boss. This is when getting your ego put away will help you since you are also their friend. A boss will emerge on emergency scenes.

It's all about having that "switch". You must know when to have fun and when to switch it off and be all business. Station life can and should be fun. Emergency scenes are where our alpha sides show.

The fire scene is a war, and the fire is the enemy. You must know your job thoroughly and be ready for the fire of your career every shift. The more you are ready, the more your crew will be ready.

Follow the leader/leading by example

Your crew will mirror your actions. If you expect them to be fully bunked out, then so should you. If you expect them to be great at forcible entry, then so should you. Get it?

However you act is how your crew acts. Positive leaders have a positive crew. Crab asses have a negative crew. There's been plenty of studies in which they have traded a positive leader for a negative one and the crews formed to their leaders. Have a worthwhile, contagious attitude to make things better.

If you have a bad attitude, teamwork will suffer. Bad attitudes mean something in your life is off. It's out of whack or missing. If you don't have your shit right, you can't expect to be on your A game.

Work and home life issues intertwine and affect each other. We bring home issues to work and work issues home. When our lives are in balance, we have less stress and issues. Recognize increasing stress and find what areas you are being weak in.

News Flash: YOUR LIFE IS THE DIRECT RESULT OF THE WORK YOU PUT INTO IT.

If your bank account is low, it's because you don't hustle. If the wife is not jumping your bones, you are not dating her. If your belly hangs over your belt, it's because you are not eating right.

You need to be the catalyst with a good attitude. You can be as influential as you want. People want to be happy.

We all choose how we react to what happens to us. We can choose a negative response or a positive one.

Have you ever been cut off in traffic? Have you ever cut someone off in traffic? Why? Probably an accident, right? Why, then, do we always assume the person who cut us off is an asshole?

Everyone has shit going on in their life. Maybe the driver is heading to the hospital or maybe not. We don't know, so why waste time getting upset? Have a good attitude every day, and your crew will follow suit.

You cannot make other people change, but you can inspire. People are all on different levels as far as work ethic and happiness. They are on a scale from one to ten, ten being top notch and one being a lazy slug. You have a crew consisting of five people: a seven, a five, a six, and eight, and four.

Your goal is to raise all of them up as much as they will let you. You cannot expect a four to be a ten unless they want to be one.

Too often we see ourselves as failures if we can't "fix" someone.

I fell victim to this thinking. I was shaken out of it when one of my guys told me that just because it's not working with one guy doesn't mean what you're doing isn't working. Look at the big picture. If the majority of your crew has excelled under your leadership, then you're a damn good leader. Don't sweat it if you don't get a 100%.

Our superpower is our passion

Whatever you are passionate about, you tend to excel at. Think about this for a bit. Those who love fishing are typically really good at it, spending all their free time they have on the water catching fish. Same goes for hunters, always in the woods bringing game home for the dinner table. So is it wrong to be passionate about being a firefighter?

It's a job that exists to save lives and protect property. I do not understand why everyone on the job is not passionate about what we do. It's exciting and we help people every day. We drive around in half a million dollar vehicle. Children look up to us. The community sees us as their safety net. Parents hand over their children to us for help even though we are complete strangers.

How to find your superpower

Let me ask you – what specifically are you passionate about with this job? Maybe

- forcible entry
- hose line management
- search
- medical
- special operations
- leadership

Whatever you like the most, you are probably really good at it. So guess what? What you are good at, you are probably passionate about, and that is your superpower.

With great power comes great responsibility.

I know you've heard that before. So now that you are aware of your superpower, you have a responsibility to use it.

Your job is to share your superpower with the rest of the world.

Share your ability by teaching it to others. Teach it to your

- crew
- shift
- department

You can get real wild and start teaching it all over your state or travel the country.

There's a need for this because most don't share what they do. They just keep it to themselves and complain when others don't perform like they do. This makes no sense to me. The more we all know or the more we all can do, the more our skills and talents benefit everyone.

The sooner my daughter gets potty trained by my wife and me, the sooner we don't have to change any more diapers. This may sound off topic, but most things we do in our family life relate to our work life.

Now, if you're struggling to find your superpower then

- what do you enjoy doing
- what comes easy to you while others may struggle
- start watching others in your firehouse
- ask those around you

Odds are they will know more of what you are good at than you will. People are more inclined to tell you what you're good at opposed to what you're bad at because we don't want to hurt each other's feelings. So start off by asking what those around you see you at excelling at.

Once you know what your superpower is, then you can start using it and sharing it.

Stay current with knowledge

We used to put fire out with buckets of water. We used to hold our breath while making a push to the seat of the fire. We used to do mouth-to-mouth resuscitations on people in cardiac arrest.

We don't anymore.

Why?

Because we stayed current with knowledge. We became more informed, we have less injuries, and we live longer now.

This is all from staying current and accepting change in what we were once taught.

Unless this is your first year in the fire service, odds are you have seen a lot of changes in practices over the years.

Seatbelts are big now. Fire apparatus come with alarms that sound if you're not buckled. Deconning your gear after a fire. Wait, what? No more dirty gear. WTF? That was the sign of a badass at one point. Now it's a sign of a dumbass.

We have more knowledge of carcinogens and the effects of long term exposure to them. We have to keep evolving with the times.

A great way to stay *in the know* is to start teaching. Teach for your department, teach at your local academy, or even go teach at one of the many fire conferences throughout this country.

You learn more by teaching others. Everyone shares their experiences, and there's always something to be gained.

Another opportunity to lead by example

The more you're seen teaching, the more those around you will teach. Truly leading by example is a great catalyst to making things better.

People model those they look up to. Imagine a crew that teaches each other on a daily basis.

I've got that.

From the new guy to the senior guy, we learn something every day. I learned more from my crew my first year as an officer then. They learned from me. It was a great experience.

Share information with crew

Always share the new knowledge you gain. There's no point keeping it bottled up in that noggin of yours. We too often assume others know what we know.

If they do, great, now they heard it again. If they don't, they've learned something new.

I was shocked at an extrication drill when I brought up using the adze end of a haligan bar to pry an opening between a car's hood and the side panel. I was taught this by my mentor when I was hired. It's a quick way to get water on an engine compartment fire before you pop the hood.

No one knew this technique. So I share it. Often...

Work more effectively

Crews can start working more effectively when they adopt best practices early on. Too often whatever we learned in the academy is what we use today. I'm not saying all past techniques do not still work. There are some better ones out there though.

In the academy, you are taught just enough to get yourself hurt. The real knowledge comes from being on the job and learning from those who constantly try to master our craft.

When did you last learn something?

Seriously, when was the last lecture, hands-on training, or class you took?

If you don't use it, you lose it.

This goes for

- humping hose
- busting doors
- throwing ladders
- saving the guy next to you

There is always some type of knowledge to be gained from every class. We must always progress in our lives with knowledge and skill sets.

Want to know why people feel down? It's because they are not progressing. They are not getting better. They are stagnant in their life. Kick negativity in the teeth, and get out there and gain some more knowledge. What will you do next?

There are classes all over the country available to take. (You can start with something simple and join my private Facebook group. You can gain access by going to www.brotherhoodcoaching.com. I post challenges I face and share the knowledge I have gained over the years.)

Knowledge is useless unless acted upon

If you just sit around and eat all day, what is going to happen? You're going to get fat. You need to get off your ass and move. The same goes with the knowledge you have in your head from on the job experiences or classes you have taken. Put what you learn into practice.

As Stephen Covey says, *we judge ourselves by our intentions – others judge us by our actions.*

Do you have your Fire Officer 45, Instructor 20, or a super graduate or doctorate degree? Great, you took classes and did the work for a piece of paper.

- what are you doing with that knowledge though
- what have you accomplished by using what you learned

Show me results on what you did – not how many pieces of paper you have on your wall. Now, there's nothing wrong with higher education and certification classes. However, I do believe too much emphasis is put on the piece of paper and not enough of on what results the person is producing.

Show me what you can do. *Talk is cheap.*

Aaron Fields – a firefighter who changed the fire service

Please tell me you've heard of Aaron Fields. If you haven't, stop reading right now and go to his site: https://nozzleforward.com. Aaron teaches a program called "The Nozzle Forward." He is one of *the best instructors* I have had the pleasure to learn from.

Aaron went out on his own to learn from some of the greats in the fire service. He

- took what he learned and twisted and tweaked a few things
- tried to break down the basics to their root and reconstruct a better method
- started by training alone on the bay floor of his firehouse
- drew interest from his crew one guy at a time

The crew was working together on mastering their skills as an engine company.

During a department wide training, Aaron's engine company showed up to perform the drill. They performed significantly better than any other crew. Now he caught the attention of his training division. They wanted to know what his crew was doing.

He shared his insights and soon started the change within his department with support from chiefs, divisions, and all the way down.

Aaron never intended to teach.

Aaron was developing his skills to better serve the community, his family, and himself. Teaching was asked of him.

Aaron teaches throughout the country with his training cadre. This cadre has shared their skill sets with firefighters all around. Fire departments have adopted their techniques in an effort to be better.

Aaron Fields is a firefighter who rides backwards on the fire engine. He does not have a white shirt nor a gold badge. Aaron Fields has made the fire service better than he found it.

> *The only shortcut is work.*
> *Aaron Fields*

Change starts with action and influence.

Too many times people think that they have to be a chief to make any type of changes. Real change starts with us – the ones on the floor. We are the ones doing the work. We are the ones who see the need. Boots on the street need to be making decisions not butts in a seat

If those boots on the street are not making the decision then those butts in the seats will.

So get involved.

Educate yourself on best practices, and influence those around you.

The goal with change is to get people to want to do what you want them to do. This is accomplished with influence, not authority.

It takes work, but I promise you, it's worth it.

Summary
If not you, then who

1. The key ingredient is to have the correct mindset – an ownership mindset will get you far in life.

2. Frank Viscuso puts it this way: If you can't lead one, you can't lead any, but if you can lead one, you can lead many.

3. To be a leader, you have to start with yourself. Get yourself squared away and show your crew the path to follow. Everyone wants to be lead; they are just waiting for someone to stand out and show them the way.

4. Egos eat brains. Keep yours in check.

5. You are the hero that you have been waiting for to show up.

6. Raise those around you to be successful.

7. We are judged by our actions, not our intentions.

8. Your passion is your superpower; share yours and learn others.

9. Get into the job. The more you put in, the more you will get out.

Now you know more about working on yourself, let's look at working together as a team.

Strategy 2
Get everyone on same page

A team in the firehouse is a team on the fire ground. This is accomplished by everyone knowing what the plays are. We must know what we must do and what the person next to us is going to do. The key points in this chapter are deciding on how things are done, encouraging team work by collaborating, and how to encourage crew to supporting the firehouse's rules.

Operational expectations

Share what you want them to do in common firefighter scenarios.

Do you watch sports? I don't. However, if you've got a YouTube video of a ripping fire with some first due companies working, I'll make the popcorn and pull up a chair.

Most firefighters love watching sports, so I like to use this example.

Professional athletes perform so well because they know what each other is doing. They have a game plan developed by the coach.

In practice (training), the offensive line (firefighters) practice their plays called by the quarterback (company officer) under the watchful eye of their coach (battalion chief).

See what I did there? *Is it clicking yet?*

The coach may also be the company officer or an admin chief at times.

If you have a younger crew with not much experience, it's best to tell them the plays at first and let the plays evolve after a lot of practice.

If you have a senior crew, have them develop the plays. Utilize the knowledge of everyone on your crew. This will help everyone feel valued by being involved in the process.

Some examples are

- a vehicle accident (extrication)
- fire (structure fire, fire alarm, car fire)
- medical calls

Below are my rules of engagement for my crew. They have evolved over time with input from my crew. This helps everyone start on the right foot.

There is no guessing.

There's a lot less ordering done on the scene when everyone already knows what is expected of them.

DAILY SCHEDULE OUTLINE	
By 0700 AM	"Operationally ready" (PPE on the truck, PAR tags in place, SCBA checked)
0700-0730	Apparatus/medical check
0730-0800	Morning meeting/Fitness
0800-1000	Training
1000-1200	House duties, groceries, etc
1200-1400	Lunch/Stress Reduction
1400-???	Hydrants, House duties, reports, target solutions, study etc.
0630 AM	AM duties

This isn't written in stone. It's just a guideline for our day. Our schedule will ultimately dictate our day. When the daily schedule is accomplished, so are we. I prefer any napping to be done in the bunk area during work day.

(The next four pages show how I organise this.)

EXPECTATIONS

- Truck equipment check
- Truck checks complete by 0730 if possible
- Engineer-assigned apparatus and equipment
- Medical-ff

We will ride a clean truck. If in doubt, let's clean it.

STATION DUTIES

- Refer to posted station duties sign in kitchen.

PPE

- Fire – (Structure Fire, Car Fire, Fire Alarm) Full turnout including SCBA.
- Engine engineer-bunker pants, if not assigned to pump the engine – Full turnout and SCBA quickly and join crew
- MVC – Bunker pants and vest. (Extrication or heavy damage wear helmet and bunker coat.)

REPORTS

If we did it, document it. Prove what we are doing in the reports. The more we document the better it is for our us and our staffing.

You don't have to stay up to put in night calls, just get up and get them in by 0630hrs so you can still help with AM duties.

CALLS

- We will answer calls promptly when dispatched.
- Seat belts are **MANDATORY** and there is **ZERO TOLERANCE** for this.
- (Drivers wait for crew to have seat belts on before moving/ FF if you're not belted and we're moving, say something)

ROLES ON CALLS

EMS CALL RESPONSIBILITY

- CAPT – Gathers info
- ENG – Pt Care Lead or get vitals/treatment (Get info if Capt. gets involved in Pt Care)
- FF – Pt Care Lead or get vitals/treatment

MVCS

- CAPT – Scene size up, call for resources assist w/ Pt Care
- ENG – Place truck behind scene to block traffic. Lights if at night, set up safe area with cones, check car for hazards, DC battery if needed and crib car, pull jump line and extrication if needed. Assist w/ Pt Care. Gather info.
- FF – Triage, report back to CAPT and then go to worst PT. If pin in, set up tools, tool man. Clean up hazards last.

FIRES

Attack

- CAPT – 360, backup (TIC, drop irons at door)
- ENG – Pump charge line with tank water, get kinks out. Possible next actions: (Feed hose from entrance door, direct water supply, set fan, collect par tags.)
- FF1 – Nozzle to front with first coupling if possible (this allows for an easier 50' hose advancement), "water the grass": assures adequate pressure
- RESCUE ENG – Hose line management, primary search
- RESCUE FF – Hose line management, primary search

SEARCH

- CAPT – Search team (TIC)
- ENG – Water supply (If needed), if not needed then join crew with 6' hook.
- FF – Search team (Irons)

RIT

- CAPT – 360, utilities, throw ladders if needed. Identify trouble
- ENG – Obtain RIT pack, gather tools (saw, sledge, haligan) Wait for others at front and monitor radio
- FF1 – Irons, 360 with CAPT- throw ladders if needed, identify trouble

INVESTIGATE

(Fire alarm or fire w/ nothing showing on arrival)

- CAPT-360 (TIC)
- ENG- FACP, ID hose line placement and water supply; Prepare to direct in units if needed.
- FF1-Investigate with CAPT (Can)

  4 person crew – nozzle and irons will be split and designated. Nozzle takes can if no line is pulled.

TRAINING

We will continually train on our basics. Our goal is to have them be second nature so that we can focus on the variable of the incident.

We will try out new stuff to see if it is applicable to our crew or FMFD.

If there is something you want or need to train on let me know and we will make it a priority. Chances are we all need work on it.

NEXT SHIFT EMAIL

I will try my best to send our next shifts "daily agenda" before that day. Sometimes our day will get screwed up and we will have to put off what we planned. Other times I may get pulled away. The next-shift email is intended to allow my crew to take care of stuff even if I am not there that way we are not on hold until I get back.

GENERAL

Remember what I told you to do when we went over our crew expectations. If we're going to do something different, I'll tell you.

Take five seconds while you are putting on your air pack and (size up the incident) for yourself. Think about what you are seeing and anticipate any safety issues and what I'm going to need you to do. This is what separates good firefighters from great ones.

Don't just be on the job.... be into the job.

Thank you.

Captain John Lovato

Be sure to sit down with everyone on your crew and go over these as soon as they get assigned to you. The sooner you do it, the less mishaps you'll have on scene.

Operational expectations make you look like pros on Super Bowl Sunday instead of a bunch of weekend warriors playing a pick-up game in the backyard.

After explaining, practice

It's not enough just to say what you're going to do. You must put the plays into action during training. We retain very little of what we are told. We retain a little more if we see it. We retain the majority of it when we do it.

Assessing performance

During your training sessions is when you want to see how everyone is doing. Not on the fire ground. Sweat on the training ground is better than blood on the fire ground. You also get to see if what you planned is understood.

If everyone is performing their assigned tasks to your level of expectation, then you know they understand what is expected of them.

The minute you start assuming is when everything will go to shit.

Firehouse principles

It is more common to see operational expectations being set up by company officers than it may be to see principles to follow in the firehouse.

These principals give each crew member the opportunity to give their input on what they like or want to have in a company officer.

Ten for you, ten for me

The ten for you, ten for me was taught to me by Mark Vonappen of *Fully Involved*. Mark comes from a football background. His father used these principles with his players. Have your crew define the principles.

You can change the principles to whatever works best. The fact that your crew defines the principles themselves is where the learning and buy in happens.

This is their opportunity to open up about

- what they want
- what they have experienced and did not like

It gives you a good insight into the person defining the principles

I save the ten for you, ten for me sheets that my guys complete. I review them every now and then to make sure I'm staying on track.

WHAT TO EXPECT FROM ONE ANOTHER

From your captain

- consistency
- sense of urgency
- never satisfied
- leadership and direction
- forthrightness
- open dialogue
- accountability
- technical command
- respect
- sense of humor

From firefighters and engineers

- sense of urgency
- concentration
- full compliance
- will to prepare
- accountability
- commitment
- willingness to play a role
- officer leads - you follow
- finish
- standard of performance (big4)

Do your job. Treat people right. Give all-out effort. Have an all in attitude.

TIP: Don't have too many expectations. Your goal is to keep these short and sweet. Too much and your people will feel overwhelmed. You want these to mean something to them, so concentrate on the most important principles to you.

Communication side effects

This ten for you, ten for me technique is a great method to open up communication amongst the crew. This interaction is typically done one on one. Feedback and clarification can be given by both parties. Clear communication prevents crappy commandos.

Don't be a rule changer to win the game

Ever start playing a game with someone who was not clear on all the rules? Did the rules seem to change throughout the game to benefit the person who came up with the game? Does this ever happen in your firehouse?

- whatever rules you set apply to you as well – if you do not follow them, they will notice and eventually call you out
- of you don't want to follow them then don't make them – it's that simple
- if you want you people in gear on a scene – then so should you

I tell my guys to look at me when in doubt. Don't have on less gear than me and you'll be fine. This lack of gear has been more and more common with chief officers on scenes throughout the fire service.

I don't understand this choice, but then again I'm not a chief so I guess I can't judge until I'm in that spot. Either way, the shoes I'm in now, I believe that if you want your people in gear, then you should be showing them by wearing it yourself.

Stay consistent so people know what to expect

Got a case of the Mondays? Cool, stay home then. Nothing will break up a crew more than inconsistency. People want to know how to act around someone

- if you're cool one day and a prick the next, they will always be on eggshells and your relationship with them will suffer
- if you're harping on one guy over his uniform then you better make sure no one else on the crew is doing the same thing and letting it go unnoticed

If a rule's not working – review it

There will be a time when expectations or your rules of engagement need to be tweaked. If someone is having an issue following one of them

- talk to them about it
- sit them down and find out what is the obstacle
- do they have a better way
- make sure it was clearly understood

You want to welcome input and feedback. Everyone has something to offer the team. Be flexible and adjust along the way if a better way is found.

Explain how to review rules

Since you encourage review feedback, make sure the crew knows how to go about it. They probably are not used to such open communication. When issues build, tempers turn up. Stress the importance of coming to you in private if there is an issue building.

I'll share with you an experience I had with a couple of guys on my crew.

"WTF, CAPT"

We were having a discussion in which one was unhappy with the results of an interaction with a chief where a class got denied.

I explained that we didn't know what the chief's reasoning was at the time and there may have been some other factors. Well, apparently I must consistently, blindly defend certain actions by our administrators.

One of the more senior members stated, "WTF, Capt! You are our captain, not Chief XXX, captain".

I'm not gonna lie. I was embarrassed, angry, and my ego was bruised. I chose not to react. I did jokingly state that the person could spend more time assigned to our busiest unit and left the room shortly after.

This interaction stayed on my mind until the next shift. I called the senior member into my office. He sat down and before I could get a word out, he apologized for the way he spoke to me the shift prior.

I thanked him for his honest feedback and input. I did ask him to make it more professional and take me aside the next time. He agreed and our relationship grew stronger after this interaction.

I knew of another officer that reacted differently after a similar situation. This officer ended all relationship building with his crew. I did not want to do this. I gained some great leadership insight from this encounter. I'm here to look after my people.

He was right, I am their captain.

I'm still a representative of the chief in a way, but sometimes your people just need to be listened to.

Get in the habit of understanding the whole story.

Evolve with crew input

Gather input from your crew. Be a caveman, a hunter and gatherer of input from your crew.

Ask what they like.

Whatever they do not mention is probably what they do not like.

People are hesitant to share negative feedback to others they have a good relationship with.

The more say people have in their environment, the more engaged they will be. No one likes to be told what to do all day long.

Make good changes permanent

If positive changes are going to be made then make sure everyone is aware of them. If you want everyone on the same page, then you must make sure they are aware of any changes.

It's like coming home to find out your wife moved the dishes to different cabinets and now you spend the morning trying to relearn where everything is.

No one likes to be blindsided.

When a rule is broken, help understand why it's broken

- show an example of it not working
- suggest an alternative
- demonstrate that alternative

Rules break. It happens. We all do it. Some of us have better reasons than others, and some of us are just dicks.

I'm not going to delve too much into conflict resolutions since that could be its own book. If an expectation or rule is not met, find out what prevented it. Get a better understanding of what occurred.

I always ask for examples so I can understand a situation better.

Ask the person for an alternative to the expectation to determine if they have a valid reason for not meeting it. It also helps if they demonstrate it to you so you both know they are capable of it. This also stresses the importance of it to you.

The majority of the time the person is either unaware or unable.

Take this approach the next time you have an unmet expectation. It's our natural tendency to assume the person is always being unwilling.

People are either unaware, unable, or unwilling when an expectation is not met.
Frank Viscuso

The assumption of unwillingness is rarely true. Most times the issue can be resolved with a better understanding of the issue.

This can be seen with reporting issues. We had some reports pulled and it was noted that medical calls were being coded wrong. Some people just assumed that the report writers were being lazy. They made up this story in their head and got upset before they knew the truth.

The correct way to complete the reports was communicated. Now reports are being done the right way. Most people's reaction was that they thought they were doing them correctly. They were UNAWARE that what they were doing was not the correct way.

We all know different information and have experienced different situations. It is an unfair habit to assume everyone knows the same or has experienced the same throughout their life.

Summary
Get everyone on the same page

1. Set operational expectations. These are the plays you expect your players to perform on emergency scenes. To play as a team, we must know what everyone will be doing.

2. Station life principles help guide daily behaviors. Crew members hold each other accountable when one is slacking. It gives everyone that extra push in the right direction.

3. Play by the rules you set for your team.

4. As we evolve in life, so must our expectations and rules. Keep your people engaged by making their input matter.

We'll now look at how to foster loyalty and respect in your firehouse.

Strategy 3
Firehouse family

What's a family to you? Probably a household consisting of parents and children.

A family goes through a lot together. There are ups like the birth of a child, birthday celebrations, anniversaries, vacations, love, and just spending time together. There are also downs that happen like money stress, deaths, sicknesses, and injuries.

What do we do on a daily basis as a crew

- we show up every shift to live under the same roof for 24 hours
- we clean and maintain our home
- we may deliver a baby, help out the injured, or fight a house fire that day
- we break bread together, share stories, and bond

There are similarities to being a family at home and a family in your firehouse.

There are people who have a happy home environment, while others have a family environment like a Jerry Springer episode.

The same can go for your firehouse as well.

This chapter will focus on encouraging your crew to work together. We will cover topics on avoiding conflict, encouraging brotherhood, and developing a family atmosphere in your firehouse.

Rollercoaster of emotions

I touched on being consistent in the last chapter. Now I'll dive in a little more.

Have you noticed that people with "Jekyll and Hyde" personalities are difficult to predict, trust and work with? Let's face it, we are all human and subject to different emotions throughout the day.

Some of these emotions are influenced by our past or whatever is going on at home. It can be challenging, but you must strive to leave your home baggage at home and your work baggage at work. My wife said it best recently: "Remember, it's just work". I'm passionate about what I do, so this is a struggle for me at times.

People like to be around happy people. They want to know that John is coming to work, not Jake, his alternative grumpy alter ego. Remember, we all have our own fights going on inside. Shit happens to all of us. We do, however, choose our reactions to them.

I've been yelled at on scene in front of other agencies over an empty bottle of peroxide. I know of people who have been yelled at over getting the wrong kind of bread at a grocery store. The wrong kind of ice cream. Some have even been called a liar over a soap dispenser.

True stories. You can't make this up. Seriously, who cares? These are all examples of someone's bottled up issues that showed their ugly heads.

Take some deep breaths, go to your happy place, and get ahold of all your craziness. We all have it.

"Your Crazy is Showing…You might want to tuck that shit back in."

We all make up our own stories in our heads on why certain things occurred. Most of the time these stories are way off. So if you've got some shit going on, talk it over with the person. Control your outburst and be hyper cool.

Let me share an experience in which people created a story in their head and got all worked up for no reason.

"I DON'T WANNA"

This event happened when we started performing a yearly combat challenge to make everyone aware of their fitness levels.

I had just returned from an IAFF peer fitness course.

This course stressed the importance of working with each other to increase fitness. By no means was it meant to push people off the job. I mean, come on, our union developed the course.

Needless to say, some people freaked out.

There were no repercussions though. Time was kept, but that feedback was just for the participant. Shoot, some even refused to do it.

They ended up having to explain themselves to the chief on why they refused to train.

To my knowledge there was no punishment though. Just encouragement to perform the drill.

It has taken time, but most understand that the drill is to make them aware of their capabilities. It is a way to encourage fitness. The drill identified individuals who have blood pressure issues and it made them aware so they were able to get treatment early.

Reactions like these seem to happen when someone feels threatened. They feel threatened because a challenge has been presented that they have not prepared themselves for.

It's important to become more harmonious. It smooths out the rollercoaster. When someone gets all worked up and erratic because they chose to react instead of understand a situation, I picture a monkey throwing poop.

"THE MONKEY IN THE BASEMENT"

My grandparents had monkeys when my mom was a kid. My grandpa even took in a monkey who was abused. The abused monkey was kept in the basement. Every time my grandma would go downstairs to do laundry, she would have to hold a clothes basket over her head. The monkey would always throw his poop at her when she appeared.

The monkey didn't understand the situation and just reacted the only way he knew how, with his very own poop grenades. Don't be a monkey throwing poop.

Great, by now you realize that you are human and have uncontrollable emotions sometimes. Get your head right and stay on a smooth ride. How do you do this

- get a massage
- run
- bike ride
- swim
- do yoga
- do burpees
- pick things up and put them down (lift weights)

All these things help relieve stress and release happy hormones.

If you're really dedicated, check out meditation. There are so many forms of meditation and a ridiculous amount of research showing the benefits. I've been meditating for over two years now. It's just like working out; the more you do it, the better you become. It helps clear your head and gets you in a relaxed state of mind.

Absent father syndrome

What happens when a child grows up lacking either a mother or a father? The child will lack the valuable influence provided by either the mother or father.

This also goes for our firehouse. The leader is the parent.

I need to be a father figure for my crew. This is not meant to be sexist. I'm a guy, so I'm going to talk from my perspective. You can replace my use of "he" with "she", my "father" with "mother", and anything else you'd like because I want the message to be understood, not the fact I wrote this from a male perspective.

Don't be the "office" officer

Ah, your own office. How sweet it is. An office has its purposes. Departments will vary on the administrative work required of an officer. However, some spend too much damn time in them watching TV or surfing the web. A company officer is the lead firefighter. You still ride a fire apparatus, go into burning buildings, sweat, and get dirty.

The more time you spend on the bay floor working with your people the better off everyone will be.

Nobody ever got better being a firefighter by being on a computer all day. Sweat equity is what you need. You want to be in an office all day? Go be a chief and deal with their headaches all day. I'm sure any one of them would trade with you in a heartbeat.

Eat with the others

I've never experienced this first hand, but I know "eating alone" happens. What kind of yahoo doesn't eat with their crew? I've had some guys bring their own food in for financial reasons or diet. I'm ok with that but I tell them they still need to be at the table with us.

A family that eats together stays together.

Chat with your crew

If your crew only sees you on calls and at meal times, then you're dropping the fucking ball. You can learn a lot about a person by spending some time getting to know them on a personal level. You are with each other every third day. The more you interact, the more enjoyable the day will be.

The less the leader is around, the more the crew will make up what to do on their own. Leaders lead. Show them what to do with your actions. Some crews may make the best of the situation, but most will just hang out. Your absence will create a monotonous environment. Think about the busy parent who spends little to no time with their child. How do they pan out?

Show up for your crew

Involved parents produce well rounded kids. Your time with your crew is what will make them better

- interact as much as you can
- be present for them
- help them with accomplishing their goals of becoming a driver or officer
- show them how to force a door better
- show them fire behavior with videos or make a dollhouse prop
- practice hose line management in the station
- set up obstacles for them to pull hose around
- pallets are free and their use is only limited by your imagination

It doesn't matter what you do – just show up!

Quit being a nagging spouse

This can be a pitfall for new officers. You are nervous at first how to interact with your crew. So you keep your distance. You do, however, start seeing issues arise, so you do what you think is right and talk to the person about it. There is nothing wrong with this. Just make sure your only interaction with someone is not telling them something negative all the time. (The clue's in the title, I call this the nagging spouse syndrome.)

We have all experienced this and probably been one before. A nagging spouse spends all day telling you what you did wrong and giving you more things to do.

"Sounds fantastic."

"Sign me up for that marriage."

I'm kidding. No one likes being on the receiving end of this, so don't do it!

There's truth to the saying of sandwiching a negative between two positives.

The typical practice involves giving a person a positive compliment, followed by the corrective recommended action, then completed with another positive.

Does it sound too rehearsed and unnatural? Try spending your day giving positive feedback.

If someone does something good, thank them. People like to feel good. They like to feel valued. Give out gifts if something big was accomplished. People don't really care about the gifts, but they do care about the act of recognizing them. It takes thought and effort to give something. That's what really counts. If you spend the majority of your time using positive reinforcement, then when you have to discuss something that needs correction, the person is more receptive.

Promote what you love instead of bashing what you hate.

Think of it like food. Spend the majority of your time eating healthy and when you have some pizza or burger once in a while, you won't get fat. The damage is done when you eat a burger or pizza every day, just like giving negative comments.

Keep it simple stupid

Long lists of to do items are overwhelming. They tend to stagger productivity more than they increase it. Keep to do lists to three to five items max. Focus on the most important tasks first, then fill in from there.

There's a lot of productivity methods to choose from. You can spend more time learning them or just take my simple advice and keep the daily list short and important.

Everybody matters

Understand where your people are coming from. We all have had different experiences in our lives. Our experiences have shaped who we are today. There's a good chance if you had the same experience they have had, you would be acting the same. I'm not saying that it's ok to be an ass clown, just be aware of what someone might have gone through or is going through. Sometimes we all need someone to just help us up.

Remember words can hurt

Words hurt, especially mean words said in public. If you want to make an enemy ridicule them in public. Here's a repeat of what I'm sure you've heard before.

Praise in public and punish in private.

Do you like to be talked to like you're stupid? Even a stupid person doesn't want to be reminded they are stupid.

The way you say things can really steer a conversation and a relationship.

People will forgive you but they won't forget.

Earn respect

Remember, people work with you, not for you. No one alone is as strong as many. Be around to help out where you can; your crew will notice. Show them they work with you and not for you.

Here are some suggestions for doing that.

Value everyone's opinions

We all see things a little bit different. Take the time to listen to other perspectives. Just because someone has a year's on-the-job experience doesn't mean their idea is not as good as someone who has been there 20 years.

Demonstrate putting ideas into practice

Listen and try the ideas out. If you only listen and then never try anything recommended, then soon no one will have input. A prime example is when we were working on implementing a courtyard lay on my engine. This was explained thoroughly in an example under "check your ego at the door".

It's like the game *Duck Hunt.* You are the hunter and the ideas from your crew are the ducks. The more often the ducks are shot down, the less the ducks will fly in front of you

- show me if it will work or show me that it won't work
- save the theories and what ifs for somebody else
- adopt the thought process of demonstrating actions to find their benefits and you'll be off to a good start
- if someone does a brilliant job, give them all the credit for it

Don't let a good job go unnoticed. Brag about the job well done. Talk about it

- in the morning meeting
- dinner time
- mention it to department heads

This will increase the likelihood of a repeat performance.

Change your words to change your relationship

When you change the words you use to communicate with someone, you can change your relationship with them. It can be that simple. Stay with me and this will all make sense shortly.

EXERCISE

Think of a person you have a close relationship with. This can be someone on your crew or a close friend. The point here is to take note of what you talk about and how you talk to each other.

Assess how you talk to them.

Odds are you talk about personal things. What they did during the weekend. How their wife and kids are doing. What they have planned for the summer. Share similar thoughts on your hobbies together. Maybe plan to meet up off duty to grab some food or drinks. You have intimate knowledge of their life.

Find a person you have little to no relationship with.

Now think of how you talk to someone on your crew that you're not close with. Maybe you don't even like them.

Assess how you talk to them.

How do your conversations go?

Do you only talk when you have to? Is everything short and sweet? Is it all work and no play?

Small talk is a norm.

Use what you know from your positive communications to improve your weaker ones.

You can strength your weaker relationship with this person by changing the way you talk to them. Start talking to them like you do a close friend. This shit works. It's not a trick. What this does is increase your relationship. You cannot fake this. You have to care about the conversation and relationship. People see right through people who don't care.

Find some common ground and go from there. Get to know more about them, their family, or their hobbies. What makes them happy? It may take time but you'll build a better relationship with them just by showing sincere interest with your words. Start simple and informal in the beginning and build from there. This one on one time with each other is crucial to bettering relationships.

They will soon be reciprocating this behavior and boom. You just made another friend, ally, and supporter.

I win my enemies by making friends with them.
Abraham Lincoln

Side by side in battle

You need to take part in the work, not just be an onlooker.

Participate in the firefight

When there is a strange sound in the middle of the night, do you send your wife or child to go check it out? So why would you send anyone on your crew into a burning building and stay outside? You are the most experienced one of the crew. You can't look out for someone you are not next to. You'll be the first to recognize changing conditions inside. The engineer is already outside watching conditions from there. Work side by side with your people.

Show you are there

Your presence will put your crew at ease. Think back to when you were a kid. When were you scared? At night when you were by yourself in the dark, right? Kids usually feel safer when mom and dad is around.

I said it before and I'll say it again: you're still a firefighter, so get in there and get dirty. This is your game day.

Demonstrate the skills you have been practicing with the crew. This helps with understanding and teaches proper experience.

If you're outside, how can you tell if the hose is being advanced with ease or not?

Put in the same amount of effort in all scenarios

Crews want an officer who will work alongside them on the fire ground and in the firehouse. Many hands make work easy. Remember this when you see your crew sweating and you're not.

Reward hard work

Would you pass up a free beer? Neither would I. So don't pass up an opportunity to reward hard work. Positive reinforcement for a job well done will lead to more jobs well done.

Make it a family

Have you ever noticed that you feel closer to someone after you have experienced a crazy event together? I know I have.

If I fought a fire with you, I would feel different about our relationship afterwards. I would be closer

- we shared a bond
- we went to battle with each other and came out victorious

I'm sure there's some deep physiological thing behind this. It comes down to this: if you're going to go into a burning building with me, then we are more than coworkers, we're a family.

If you're a family on the fire ground, then you better start acting like a family around the firehouse.

I talked earlier about changing your relationship by changing your conversation.

That's one way to better your bond, and here are some more examples.

Captain Cook Sunday

I got this from a former captain of mine. On Sundays, my crew typically goes out to a restaurant for breakfast. We don't normally go out to eat so this is a special treat for us. Everyone can get what they want, and we get to interact with the public.

As far as dinner goes for the day, I prepare it. An officer has some extra duties to perform than the rest of the crew, and I've found that I have less on Sundays.

I use the extra time to cook for my crew. They handle the cooking during weekdays so this is my chance to return the favor.

What do I cook? There are never hamburgers or hotdogs cooked on Captain Cook Sunday. Some entrees have been eggplant bolognese, chicken marsala, seafood gumbo, shrimp puttanesca, and even bone in rib-eye steaks if we are celebrating someone's anniversary or birthday.

Not only do the guys like the food, but they appreciate the act of me spending the time to make them a meal.

Invest in some interactive games

Remember days before everyone was staring at screens such as TVs, smartphones and laptops? Firefighters used to play real-world games, like

- cards
- checkers
- handball
- horseshoes

Give it shot.

I found that we can get everyone involved in a game of "around the world". There's no limit to the amount of people who can play. I am completely horrible at it, but I enjoy the time together with my crew.

Find something to do actively with each other instead of staring at your phones or TV during down time.

Watch a show together

I know I just said not to stare at a screen, but I'm a big movie fan. Throw in some popcorn, and we can be friends.

I'm always down for a good movie I haven't seen; the trick is finding one I haven't seen.

This is another opportunity to hang out with each other and bond as a group.

Hold family days

You would think this would be a no-brainer for an officer to do, but the suggestion to hold the first one we had came from an engineer.

We were working Thanksgiving so he suggested that everyone invite our families up. That's exactly what we did. Extended families and all.

We ended up with over 30 people sharing a meal in the firehouse bay. My wife's side of the family chose to celebrate Thanksgiving with us as well.

My nephews were so excited and bragged to all their friends at school.

This gives you a chance to get to know the spouses and kids of the people you work with. It was a lot of work but worth it.

We have a few members who commute over two hours to work, so their families rarely visit. One of the guys' wife had the idea of surprising their husbands by showing up one holiday with the kids. She got ahold of me first and we all had our family members show up.

I may have had a little more fun with this by telling the two guys we had a firehouse tour showing up around lunch time and that they had to be in the button up shirts for the event. Our typically daily uniform is a t-shirt, so most guys loathe putting on a collared shirt. Well, the guys ended up in their collared shirts awaiting this tour right at lunch time. The look on their faces when they saw their wives and kids in the parking lot was priceless.

We all had a great time hanging out with each other and our families. Days like these are so beneficial for a crew. One of our guys who is not that social and doesn't have his family come up normally had his show up for this one. It was great to see everyone interacting.

Celebrate special events. Invite your family up to meet your firehouse family.

Be proud of your station

This is *my firehouse*. There are many like it, but this one is mine.

My firehouse is my best friend. It is my life. I must master it as I must master my life.

So the riflemen creed from full metal jacket can be tweaked for us too.

Get a station logo

Pride of the Southside is the motto on our company patch. My department has been around since 1901. Our first station patch didn't exist until 2009. There's only one way to have one if one does not already exist. That way is to make one. Some don't understand the importance of a station logo. I even overheard someone say it is harmful and will divide the department.

Let me break it down for you.

You want a prideful, engaged department?

First you need a prideful, engaged firehouse. How do you do this?

Give them an identity.

This idea started off in the military. Different units have different company patches. The same holds true for the fire service. A station patch gives the crew a unique identity within the department. Throw a station patch on an apparatus, and it will always be clean. Things get simpler when we break them down. Break down the department into individual fire stations. Then put them all together.

Most places allow firefighters to wear their station patches while on duty. This helps with the ownership mentality.

Once you have station patches, you can place them on

- shirts
- car stickers
- coffee mugs

Crews like identifying with their crew and department.

Ever meet someone in their eighties with a ratted up firefighter hat? That's one proud SOB right there. Everyone wants to belong to something special. Make your firehouse stand out by having a logo to identify with.

Visit your neighbors

Who's your second due? Reach out to them for training or meals. This is my new goal for this year. We operate pretty well together as a crew, but I rarely train with my neighboring firehouses. There's only one way to operate well together on the fire ground: train together.

Make phone call and have them meet you at the training ground. Ask them to show you how they handle being first in at a fire and show them how you normally operate and what you would like your second due to expect.

Who's the station champ

Make training more fun by turning a drill into a competition. Who can

- force a door the quickest
- mask up the fastest
- get a line to the third floor the quickest

and so on. A friendly rivalry is fun – just keep it to the drill ground and not the fire ground. We are all one team in the end.

Return on investment (ROI)

Do you have a deferred comp? Savings account? How much do you put into it? The more money you invest, the more you will get in return. This same principle applies to people. What can you do for them?

Invest in your team

Think like this: Are you more inclined to help someone who is nice to you and helps you or would you rather be told what to do by someone who is always short and treats you like a piece of equipment?

Make daily deposits in your people's emotional bank account by doing some of the activities I have recommended earlier in this book. The better you treat people, the better they will treat you. This same principle works the opposite way too. If you want to act like an ass, don't be surprised to find a shit sandwich on your plate later.

Encourage them by giving more than taking

At times you may have to give more than you take. That's the way it is sometimes. Sitting back and waiting for someone to make the first move will disappoint you in the end. Give more value than you take from others, and you will be on the road to success.

Be a keynote speaker

You need to motivate people. The fact your crew already showed up for work means they are motivated. Shocking to believe, but it's true.

We are all at different levels, and it's the officer's job to make everyone increase their level. But how? There wasn't a certification class for this.

Be the motivator

When was the last class you attended? Did you feel fired up afterwards? Have you ever been to a fire conference or FDIC in Indy? Both of these events have a keynote speaker who gets everyone fired up at the start of the events. Be a keynote speaker for your crew.

You can look on YouTube for recordings of keynote speakers and play them before your morning meeting to help get everyone's mind in the right place.

You can also have your own routine.

A keynote speaker draws the audience in and sets their mindset to be positive and excited. Do the same thing for the people in your firehouse.

Summary
Firehouse family

1. Keep your emotions in check and be the same person every shift.

2. Be present for your crew. They learn from watching you.

3. Give out more praise than you do criticism – a lot more.

4. More brains work better than one brain. Take ideas and input from crew and use it.

5. Change your words to change your relationship.

6. Fight side by side with your crew.

7. Families to who play together stay together.

8. Invest in your crew. The ROI is priceless.

9. You are your crew's keynote speaker. Act like Tony Robbins and fire them up.

When they need help they will call us. But when we need help, all we have is each other. It's a brotherhood.
Patrick Brown – FDNY 3 Truck
Killed 9/11/01 (North Tower)

Strategy 4
Your time is valuable

"I don't have time" is not an excuse. It's a statement revealing your priorities. Everyone is busy. Some are productive though. Productive people focus on what's important first. When you start valuing your own time, you will also value others' time. Spend your day living rather than just existing.

Set your priorities

It's not about having time, It's about *making* time. You need to set priorities; resources are not limitless. I set mine in this order

- firefighters
- fire apparatus
- firehouse

Our day is filled with tasks that need to be accomplished. Some tasks are proactive and others are reactive. We work in an environment where we are drawn away at a moment's notice to handle an emergency scene.

Plan your day or your day will plan you. Chief John Eversole, Chicago Fire Department, said

> *Our department makes 1,120 calls per day. Do you know how many of those calls the public expects perfection on? 1,120. Nobody ever calls the fire department and says, 'Send me two dumb-ass firefighters in a pick-up truck'. In three minutes they want five brain surgeons – decathlon champions to come out and solve all their problems.*

The family down the street whose house is burning and child is still inside does not give a shit about a dirty toilet or if the firehouse garbage was taken out. They want *badass firefighters* on scene saving their most prized possession.

Firefighters put the fires out

The firefighters on your crew are priority number one.

The public expects the best of the best to show up and help them. To be the best of the best, we must be prepared for the job we are about to encounter.

Clocking in at my station means your

- gear is set up on your assigned apparatus
- your SCBA is checked
- your par tags are in place

Expect fire with victims trapped every shift and set yourself up accordingly.

"BIG FIRE"

The biggest fire of my career came in 10 minutes before my shift started. All my gear was ready on the engine when we were dispatched to a two-story apartment fire, two buildings involved, and possible victims still inside.

At the start of each shift, I check my air pack first thing when I arrive on shift. Since 2002, I can recall at least 10 times when I would not have been able to use my airpack if we caught a fire. Either there was a regulator issue, the air bottle was not attached, or a leak was present somewhere.

For a team to be successful, everyone must be ready. You are letting your crew down if you are not making sure you are ready to help at the next fire.

Fire apparatus gets us to the fire

Now that we are ready, the apparatus comes next. We need to be able to get to the scene and use our tools.

Complete the daily checks of your

- apparatus
- equipment
- medical supplies

This should be done every day, so get it done and out of the way.

It's a known, high priority task.

I've heard of people taking all day to check their rigs. What kind of crew do you want responding on one of your loved ones?

The sooner we find out that something needs to be fixed or is missing, the sooner we can get the issue resolved.

You can make a *Plan B* if you have missing or broken equipment, but you have to know about it first.

Checking our rigs is preventative maintenance. When we check them daily, we increase the likelihood of catching issues early on.

The firehouse is the sideline to the game

The firehouse duties come after you make sure you are squared away and the apparatus is ready to go.

The firehouse is our home and we must take care of it, but it does not

- put fires out
- help save a life

If a clean toilet is more important than knowing your SCBA will work then you've got some ass-backwards priorities.

Too often I've seen people give a new guy a hard time about cleaning the bathroom or vacuuming a room, but they could care less if the guy can force a door well or maneuver a line with ease.

Worry about the important things, and the rest will fall into place.

Plan your daily agenda

Successful people plan their day. There are paper planners or planner apps for our phones (I personally use 2do).

Both create an agenda for the day. This helps you stay on track with your priorities.

Create a PDF to plan the shift

Find an agenda form online or make your own. It can be simple or complex. Even writing three things on a sheet of paper is better than nothing. Our brains are made for ideas, not storing them. You are more likely to accomplish something when you write it down.

When you spend the time to create a fillable agenda, it is easier to plan future shifts. Forms act like a trigger for your memory to make sure you address a certain area.

- a spot for training reminds you to plan training every shift
- a maintenance area will remind you of any issues passed along to you at shift change

Forms should be customized to how you think and go about your daily routine.

Email it to your supervisor

Now that you have your agenda figured out, send it to your supervisor. Keeping your supervisor informed on what you are doing is also helpful for their planning.

They are also less likely to give you busy work if they already know you are being productive.

You can save time and cut down on phone calls by keeping your supervisor in the loop of your daily activities with this simple email.

Email it to crew

I work on our daily agenda the shift before. I then email it to my crew so they have an idea of what to expect our next shift.

No one likes to be surprised or blindsided. Letting people know what is going on also enables them to be able to plan what they would like to do.

Adjust it

The morning of shift start, I review our company journal, training calendar, emails, and adjust our agenda as needed. There are times the day may go sideways on us, but planning your day reduces these occurrences.

Review with crew at morning meeting

Our morning meeting is conducted 30 to 45 minutes into the shift. By this time the apparatus and medical checks are typically complete and any phone calls from the battalion chief about last minute changes have been made.

We review riding assignments, training, and anything that has to get done. If anyone has input about anything else they'd like to do, this is when they do it.

Most of the crew has already seen the agenda in their email the shift prior, so this enables them to already have a heads up on the plans for the day.

Follow during the day

Keep track of your items on your agenda and makes notes as necessary. This helps when it comes time to make entries in your company journal. Think of your daily agenda as a meeting announcement and your entries in the company journal as the minutes to what happened in the meeting.

Using an agenda and company journal helps keep you on track and paints a picture of what happened so other shifts can pick up where you left off. This keeps communication flowing smoothly in your firehouse. When communication runs smoothly, operations run smoothly.

Pick one high priority item per day

When I first started using an agenda to follow along at work, I had it packed full of trainings, projects, and things to do. Every shift I would end up crossing things off that we would not get to. I was setting the culture that if we didn't get to something then it was ok.

I changed my strategy and began putting one high priority item on the to do list. This item would get done, even if it was late in the afternoon. Normally, it's a high value training.

My guys know I'm antsy until we accomplish some type of good training together and then I'm ok. I've explained to them that if we have not made ourselves better somehow then I feel I have failed as a leader that day.

Smaller items can follow

I do plan more than one item on our agenda, but these items can be accomplished other days if we do not get to them.

TIP: Putting too many things on your to do list can cause you to feel overwhelmed and slow productivity.

My personal to do list that I follow every day has anywhere from three to five things on it

- prioritized your items by importance
- do the most important thing first then move on
- aim to complete three tasks up to five max

This has been working very well for me. This and the fact that my daughter takes a nap for a few hours in the afternoon is actually how I'm making the time to write this book right now.

Here is the agenda I use on a daily basis in my firehouse

COMPANY 13 DAILY AGENDA

Date: 9/18/16

Operationally ready by 0700

Truck/Medical check 0700-0730

Morning meeting @ 0730

ENGINE 13	ASSIGNMENT
Officer Lovato TIC	Monthly: Hydrants 20/20
Driver ███ or a	Preplan :??
FF ███ or Nozzle/Can	
FF ███ or irons	

RESCUE 13

Driver ███ or

FF   Hook

DAYS EVENTS

Meeting: New BC announcement

WOD: Yoga

Training: Search hands on drills: primary, move victims, remove through window in gear

Things to do: Pick someone for September's preplan

Special events: Breakfast and Captain Cook Sunday

Daily log:

Focus

Nothing is gained from playing on your phone or watching TV. They are time wasters. Do I do them? Sure, but only after my big goals are done. You are actually better off sleeping then sitting in chair and watching TV, as far as relaxing goes. When you watch TV, your brain is still be stimulated by the images. Your brain rests when you are sleeping. I'll dive more into this later on.

Keep focused on your high priority items first. The way you feel after working out is similar to the way you'll feel when you check off high priority items on your list. This deals with our psychological response and a whole bunch of other studies done by people with a lot more initials after their name then me. Want to learn more, get the book *The One Thing* by Gary Keller.

Work on one thing at a time

Multitasking has been proven time and time again to be poor at efficiency. Don't believe me; keep doing it and let me know how things work out. I think of it like this: You are half-assing two things. Just whole ass one thing and move on to the next.

Relax

To keep focus you need to take a break from time to time. This enables you to push harder when you are working.

Have you ever done a tabata workout? It's where you do 20 seconds of balls out work followed by a 10 second rest. You keep repeating this cycle until you have completed the amount of rounds you wanted to do. You are able to push hard for those 20 seconds since you give your body a rest for 10 seconds. It gives you a great cardio workout. The same principle applies to working better.

Taking periodic rests is what enables us to work more efficiently.

Procrastination is avoiding success

We all have the same amount of time in a day. There's no such thing as not enough time. You just didn't care enough about the task to make it a priority to get done. It's the hard truth but we have all used this excuse of not having enough time. We use it because people accept it. Just know you are lying to yourself when you use it.

When my daughter was an infant, my firehouse had its busiest unit browned out. Sleep was a thing of the past. I made the excuse of lack of sleep for not working out. It wasn't until my pants started fitting too snug that I realized I had to quit lying to myself and change.

Hello, 4 am. Yep. That's when my day starts on my duty days. It begins with a double shot of espresso followed by some reading and meditation. It moves on to the gym where I pick things up and put them down.

I purchased an exercise bike to use on days I don't lift or have to be home since my wife works as well. Sure, my gym has a daycare, but I'm an overprotective father and I'm not a fan of my daughter playing with snot-covered toys from some other kid. She's my princess, and that's the way it is.

Now, since I choose this route, I have to be more creative with my fitness routine. The point is, if you want to truly accomplish something, you'll find a way.

Prioritize what you want to accomplish

How do we accomplish anything? We prioritize it. Guess what? Remember when I really wanted to get promoted to captain on my second promotional exam? I studied more than an Instagram model takes selfies. I put in 40 hours a week of studying. I bounced around from libraries to coffee shops. Whatever I had to do to put in the time.

Here's a little tip I learned. You want something real bad? Get a coach and put in the work.

People think that you have to be super smart to score high on promotional exams. Does it help? Sure. What it really comes down to, though, is putting in the hours of work. The super smart guy might not have to work as hard, but if you put in more work, you'll get it.

This same thinking goes for your daily life around your firehouse. Put in the work, and you'll get what you want. If you are not getting what you want, then you didn't put in enough work.

People are rewarded in public for what they've practiced in private obsessively, intensely, and relentlessly.
Tony Robbins

Prioritize the most important tasks and focus on them until they're accomplished.

Maximize your time

We all have 24 hours in day. Some just do more with theirs than others.

There's a story that goes, if I gave you $86,400 right now and told you that you can do whatever you'd like with the money. However, the deal is that I would take back any left over in 24 hours. What would you do with it? You'd spend all of it. Maybe on experiences or items, whatever the case may be, you would use it all.

Well, we all have 86,400 seconds in a day. Once it's gone, it's gone. No one has figured out how to get time back. So what are you doing with yours?

Resistance might mean it's not a priority

When working with a team of people and you are getting resistance or tasks are not being accomplished, chances are the task is not a priority for the person. I'm sure you'll hear the *we didn't have time* excuse, but it all comes down to what the person prioritizes. This is where a good leader figures out a way to get the person to want to make the items a priority.

Threatening may work in the short term, but it will have long term negative effects. They call it the art of leadership for a reason. Try something different so the person values the task more.

The following phrase is like nails on a chalk board to me: "If you don't, I'll write you up". I have never heard a great leader use this phrase. To me, this is like using the deck gun to extinguish a lit cigarette. Will there be times when paperwork is necessary? Unfortunately, yes. This, however, should not be your go to tactic for everything. People perceive this as a threat and are only doing the task out of force, not will. Congratulations, you're a manager.

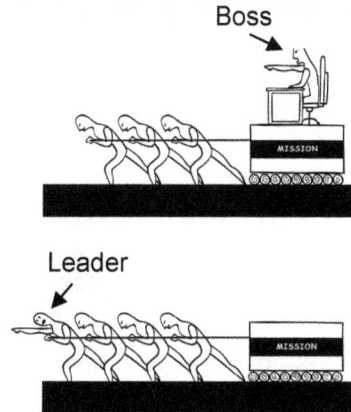

A goal without a date is just a dream

Every one dreams of success in one form or another. The ones who achieve their dreams are the ones who made it a goal. What's the difference? I'll explain.

Due dates make things happen

Goals have a date that they are due. Put a date on something and see it get accomplished.

Remember back in school and you had an assignment due in two weeks?

When did you do it? The day before it was due? We tend to put things off unless we know they have to be done by a certain time.

When we put dates for a goal, then we can break it up into smaller tasks and complete the big goal on time. The date puts a finish line to the goal you want to accomplish.

Plan better

You are able to plan better once you have a date for your goal. How do you eat an elephant? One bite at a time. Use this thinking for your goals that seem too big to accomplish

- break up the goal into smaller digestible goals
- take bites out of it every day

It will be less intimidating this way. If you want to lose 10lbs, you don't do it in one day. You do it by working out a little bit every day.

Work effectively and efficiently

Breaking up a goal into smaller tasks with dates will help you accomplish tasks easier. You'll start to find that you don't have to be busy all day to be productive. It's all about using our time better to accomplish high priory tasks first.

At a structure fire, we have to put the fire out first. Then we can do overhaul and salvage work. Once the fire goes out, everything gets better. When the high priority tasks don't get accomplished first, then your day can go to shit quickly.

Plan your game plan

You don't know when an incident will happen. So you must be prepared for when it does happen.

What has happened at every fire you have gone to

- you probably put it out with a hose or a water can
- somebody may have had to force a door.
- the building was searched for any occupants
- there was probably some overhaul work and investigation done at the end

So when do you decide to plan what you're going to do? Do you wait till you are on scene with flames shooting out of windows, civilians yelling, adrenaline pumping, which leads to a lot of shouting? Or do you practice common tasks ahead of time and assign them so everyone knows their job and what to expect?

You are either being proactive with your crew or reactive. There will be far less anxiety, mistakes, and shouting if you choose to be proactive and go over common game plans ahead of time at the firehouse.

Pre-assignments

We discussed this previously in strategy two.

Go over what you expect each position to do if you are first in at a fire and when assigned

- search
- RIT
- water supply

People are more likely to succeed when they know what is expected of them. Let's face it, the fire ground can be a stressful environment. Why make it worse by winging it until you get there?

If I have a four person engine company, I have the thermal imaging camera (TIC) and I assign the seat

- behind me is the nozzle or the water can
- behind the driver is for the irons

The two firefighters know that the high priority tools have been assigned and they can concentrate on their "one" tool. This decreases anxiety and allows them to focus.

When I started as a firefighter, we would not know who had the nozzle until we got on scene. If the fire was on your side, you had the nozzle and the other guy would grab the irons. Anxiety was higher since you did not know what you were grabbing until you got on scene of a building with flames pushing out of it. This also lead to nozzle stealing and created more of a chaotic scene rather than a well-rehearsed one.

Do what you want, but I will say that my fire scenes now go a lot smoother and there are no guessing games about where someone is or what they are doing

- who is doing what at a pin in
- who's working the spreaders, cutters, vehicle stabilization
- plan ahead of time the specific tools that you will need

A game plan helps everyone know what each other will be doing at an incident. This helps with accountability and what to train on. *25 to Survive: Reducing Residential Injury and LODD* states

You must have predetermined engine company positions in order to provide the greatest efficiency in the delivery of water to the seat of the fire.

Duty board

Do you have a duty board in your firehouse? Not only is this a traditional item but it serves a higher purpose that most are unaware of.

The duty board is a clear and prominent visual aid to your daily agenda.

Very few people can be told something once, be given distractions all day long, and still accomplish the task without some type of reminder.

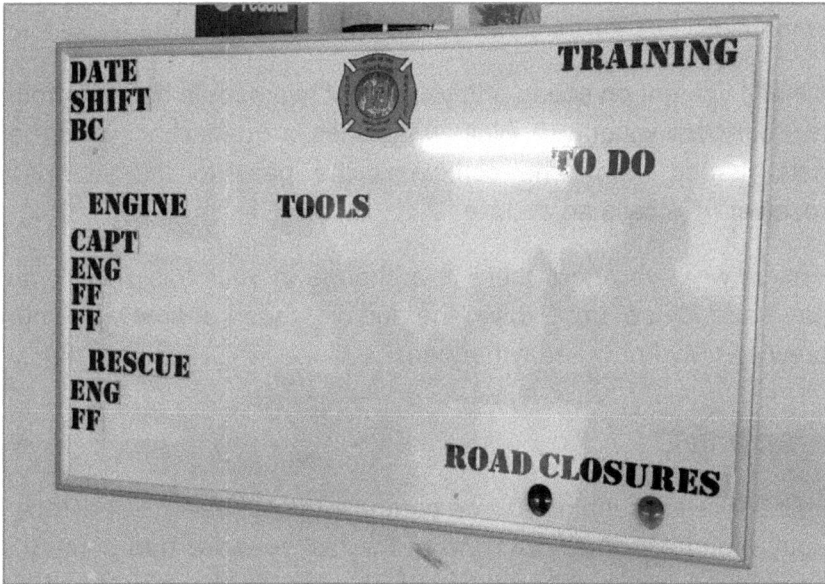

Daily plan

Your duty board should be a larger version of your daily agenda. What you have on it can vary to what your needs are.

Assignments

Put up the daily tasks. Now everyone knows what has been done or still needs to be done without being verbally or, in this case, visually reminded all day.

Positions

Placing riding positions on the board with tool assignments is a great way to provide a reminder to those on the apparatus. This is especially useful when you have someone transferred in from another station who doesn't normally operate with a game plan.

Call an audible

You've set your team up for success. You have assigned tasks ahead of time and communicated your plan so they can all operate smoothly.

What if you get on scene and you've got two people hanging from two different windows? Well, what does a quarterback do if he gets rushed before he completes the pass to the intended receiver? He calls an audible.

This is what you do if there is a change in your basic plan you have already set up. Explain the audible ahead of time with your crew so they know it may happen.

"AUDIBLE"

When we are running a three person engine, I become the irons guy along with the TIC. One of the last fires we had come in ended up being a burnt up air handler on the outside of a building. We didn't need the irons so the nozzle guy was instructed to just leave them next to the engine. It all comes back to communicating clearly.

If you see something, say something

Explain to your crew that they can also call an audible. Having a plan does not handcuff you to that task only. We all know that plans don't go 100% so we have to allow for variances. Your crew has to know how to change to accommodate the variance though without freelancing.

Check with your supervisor

I've instructed my crew that if there's a change and someone decides it's best to do something different than what was already planned for them, they must let me know before they do it.

It's not asking permission; it's keeping me informed.

I'm expecting certain tasks to be completed by certain people. When there's a deviation from the plan, I must be aware so I can adjust.

"WHERE'D HE GO?"

I normally have two units running out of my station. An engine and a non-transport rescue unit with two guys. Just like the show *Emergency*!

My rescue is assigned to

- bring a hook
- assist with hose line management
- be prepared to be search

We got called for a structure fire. It ended up being an outside shed fire.

The nozzle firefighter and I proceeded in between the two houses to the shed and extinguished it. The rescue firefighter showed up to help but with no rescue driver.

I was planning on giving them an assignment, but the crew was split up.

I talked to the driver later. He explained that he had helped the engine driver with some tasks since he figured I didn't also need him.

When it comes down to it, this is freelancing.

We spoke and I understood why he did what he did, but I also explained that I was expecting him to show up with his firefighter to help us.

He understood and I learned the value in communicating to the whole crew that if there's a change you want to make, I have to be included with the change so we can stay on the same page.

The same rule applies to myself. If I call an audible, the whole team must be aware of the change so we can be successful with our mission.

Make sure everyone understands the new plan

Be sure to get some type of feedback so everyone understands what is going to be done. Communicating involves a sender giving and a receiver understanding with some type of feedback to ensure the sender's information was understood. Shouting in random directions is not communicating.

Never assume – over-communicate

The more we repeat our message in different ways, the better the chance that we are clearly understood. People are horrible at communicating. Use different communication methods. Just because you understand the email you're sending doesn't mean that the reader will understand it the same way. Ever have someone get upset with you after a text or email and have no idea why? We judge ourselves by our intention and others judge us by our actions.

Communication method	Description
Written	My daily agenda and operational expectations are written down. Every member is given a copy of the expectations when they get assigned to my station and receive the agenda the day before.
Verbal	When a person first gets assigned to my station, I review the operational expectations with them in a one on one meeting. Our daily agenda is also verbally reviewed in our morning meeting.
Visual	I've found when you use a whiteboard or chalk board as visual aid, it helps with retention and understanding.
Practice	Now that you have got a plan and communicated it, you need to practice it with training to ensure everyone has a clear understanding and able to perform. Talk is cheap; show me what you know.

Nobody likes busy work

Have you ever worked in an environment where people hid in an effort to avoid work? That's the environment you start to create when you choose to be reactive instead of proactive. If people are consistently given work when they are seen not doing something by a supervisor then they learn not to be seen.

When our daily agenda is done, so are we. The only exceptions are calls and anything dropped in our laps last minute from the battalion chief. The crew is aware of this and likes it.

This allows them to plan anything that they want to complete, like

- workouts
- studying
- tasks

If you decide to keep finding work for people to do, they will start to drag tasks out longer than necessary to avoid additional work. Give people a finish line so they know when they are done and how much work it will take to get there.

Now you know how to set up a system to run your firehouse operations at peak performance. Next we will look at how to run your personnel at peak performance.

Summary
Your time is valuable

1. Your priorities: the firefighters, fire apparatus, then the firehouse.

2. Use a daily agenda.

3. Share with your crew and supervisor.

4. Focus on one task at a time.

5. Multitasking is half assing two things, whole ass one thing.

6. Procrastination is avoiding success.

7. Put a date on a goal if you want to achieve it.

8. Pros have a game plan, amateurs wing it.

9. Duty boards are more than tradition.

10. Adapt and overcome with an audible.

11. Constant communication helps people hold onto what matters most.

12. Being busy and being productive are two different things.

13. Focus on productive tasks.

If you are willing to do only what's easy, life will be hard. But if you are willing to do what's hard, life will be easy.
T. Harv Eker

Strategy 5
Operating at peak performance

Do you have kids who nap? How do kids act when they miss a nap or when it's close to bed time? They probably start getting crabby and irritable. They are exhausted and need rest.

We are the same way. When basic needs (like hunger and sleep) are not met, we are more irritable and less on our game.

It's important to be focused when we are performing lifesaving tasks. We make better decisions when we are not physically exhausted and hungry.

Shift work varies from department to department just like call volume and workload. Some work 8s, 10s, 12s, 24s, or the increasingly popular 48s. Time is of the essence when an emergency call comes in, so how do we keep our people skilled and able to perform at their best

- if we grind all day then we will be less effective the rest of the shift
- if we rest all day then our skills will go to shit

We have to keep a healthy balance of work and rest, like the tabata workout I described earlier where we give it our all for 20 seconds, then rest for 10 seconds. We just have to be creative.

Here is a way I have achieved this balance.

Stress-reduction time

Let's reset the clock on stress. Many organizations have seen the value in building in rest during the day for a long time. Companies like Zappos and Google have actually built nap rooms into their offices. They have found that naps have boosted productivity.

Different strokes for different folks

The morning workload will determine the amount of stress reduction I need in the afternoon. Let me make this clear for anyone who just drank a big glass of "hateraid": My crew is not behind on anything.

We are the busiest station in the department (5,000 runs). We are ahead in our training hours. Our assigned work for the department gets done every month, our daily chores get done every day, and we produce results on scene.

Our shift starts at 0700 hrs. The majority of my crew is up by 0400hrs. A few commute over 2 hours to get to work or get a good workout in before shift. So by lunch time, most of us have been chipping away at the day for eight hours already.

There are days when either a 20 minute nap or meditation does the trick. Other days may need longer if we have been out in the Florida sun all morning doing work.

Choose what works best for you

Some may choose to relax in a Lazy Boy or their bed. I will tell you that to get the most bang for your buck, hit the bed, turn off the lights, and close your eyes. Believe it or not, sitting in a Lazy Boy and watching TV does not give you the same recharge as climbing into bed and closing your eyes.

Sprinter versus long distance runner

We need to be ready for action at a moment's notice. Years ago, I worked with a guy who had just finished up running five miles on a treadmill. Just as he finished, he was dispatched for a fire in our downtown area. This fire ending up going to multi alarms.

How do you think he performed?

He was already wiped from his workout and now had to keep grinding through at this fire. He learned a valuable lesson of balancing his workouts at work. This goes into how we balance our days as well.

You can be a sprinter or a long distance runner. Both runners have the ability, but the long distance runner has a long, difficult session while the sprinter has a shorter, all-out burst of effort followed by rest.

Grinding all day long

Grinding all day long is just like the long distance runner. You can get through it, but it will take a toll on your body. You cannot sustain muscle without appropriate rest. Will you be worth a shit at the 2 a.m. fire if you stayed busy all day just for the sake of staying busy?

Work/Rest balance

Follow workloads with appropriate rest. Have you ever taken an all-day course before? How long before your attention span faded and you stopped taking in the information?

Some of the best teachers build in breaks to keep their listeners engaged. This same strategy applies to the workplace.

You can't give 100% unless you're at 100%

Our performance is undermined when we are

- physically tired
- mentally tired
- emotionally exhausted

Mistakes are more likely to happen when we are fatigued. To maintain at optimal levels at our firehouse, we have instituted stress reduction time. Our stress reduction time gets built around our daily work load. High priority items get attention first during the morning hours, and we get to the rest around our stress reduction time. The only things that interrupt stress reduction are emergency calls. We are all aware that stress reduction activities may be shorter on some days but it will be there if we can make it happen. This goes a long way.

Stress reduction is a valuable tool when used correctly. I had a period in my career when stress reduction was not allowed. I remember days when I was ready to go home at 1900 hrs or having issues driving to night calls since I was just exhausted. Why do we do this to ourselves? Big corporations recognize the value in giving their people some time to rest. I'd rather be productive than busy.

Think of stress reduction time as Stephen Covey's 7th habit, *sharpen the saw.* You keep yourself fresh and engaged when you sharpen the saw.

Give me six hours to chop down a tree and
I will spend the first four sharpening the
axe.
Abraham Lincoln

Next we will look at how to bring out the best in your crew using simple morale-boosting techniques

Summary
Operating at peak performance

1. It's important to stay fresh to operate at your best. Not only will a built-in stress reduction time benefit you, but your crew will appreciate it as well.

2. Implement some time for stress reduction.

3. Continue to operate at 100% by taking the time to "sharpen your saw".

The secret to becoming more productive is
not managing your time but your energy.
Michael Hyatt

Strategy 6
The secret to increased morale

Have you ever see a happier crew than one who just caught a job? Doing our job makes us happy. We feel value when we perform the skills we specialized in. Companies who catch fire regularly start seeing a decline in morale after a few shifts without a fire. When was your last fire? A week, month, year ago? How can we keep people motivated and committed to the team with long spans between catching a fire? We can replicate the satisfaction felt after performing well at a structure fire with regular, high valued, quality training.

If you're about to say that all training has high value, then I'm going to disagree with you. Training can be categorized into different value levels, and the value of it will be placed by the student. When people attend your class or perform the training assigned, then they view it as valuable.

After being on the job for over 10 years, I was assigned training on how to remove medical gloves properly. True story, I was there. How valuable do you think I saw this training as? Does this training have a need? Sure, in EMT school.

How about some fire training in which you go out to an empty parking lot and pull a hose 200' in one direction and flow water off into the distance.

- did you learn how to estimate your stretch
- did you have a target to stretch to
- how big was the fire you needed to attack
- did you practice hose line management
- how did you move your nozzle to direct your stream

Examples like the ones above are why people don't like to train. It's human nature to not want to do something that we see little value in. To change this mindset, we must change how we deliver training and the content we are using.

Here's another quick story from a senior member I used to work with. He was not too keen on training every day until he saw that the training we were doing was relevant to what our job entailed. He shared that he was once forced to watch an ice rescue training video years back. Our winter temperatures rarely drop below 70 degrees.

Fire conferences exist for a reason

Fire conferences and seminars are everywhere now. So are private training groups composed of motivated firefighters who travel around sharing their knowledge and skill sets. Businesses do not succeed unless there is a need for them. So why are these events so popular?

Poor quality departmental training

People go outside of their department when their needs are not being met. I have yet to see someone at a fire conference or training event that has a negative attitude or is happy with the status quo. The ones at these events want to be challenge and to become better firefighters.

Why do people stray from their relationships? Their needs are not being met. Why do kids act out in the classroom? They are not being challenged. Can you relate to these things in your workplace?

Training takes effort and consumes resources

Quality training takes work. Most training divisions have limited resources or are nonexistent. When money gets tight, what's first to be cut? Training. Fire conferences and instructors are not rolling in the dough. The majority of the costs go to props. Some of the instructors even volunteer their time at these events just to keep the costs down for the students. These instructors are essentially the "senior men" the attending firefighters never had.

Take this concept of fire conferences and apply it to your firehouse. Firefighters want to flow hand lines while advancing through a building. Firefighters want to force doors with their irons and saws. Firefighters want to throw ladders to buildings to rescue occupants. Firefighters want to practice their skill sets in the most realistic conditions as possible.

The training formula

When we train we build skill, as skill increases we build confidence. An increase in confidence decreases performance anxiety. All of these factors put to gather increase morale.

In other words

Training = Increased Morale

Not all training is the same, so let's talk strategies about how to deliver the best training we can.

Train together

When you train together you

- build skills
- build trust
- build confidence
- reduce anxiety
- become more effective as a team

Firefighting is a team sport. Every role needs the others in order to have a successful outcome. In order to be successful on the fire ground, we must sweat together on the training ground. A probie performing tasks while everyone watches does not prepare the team. If this is extra training to help them get up to speed, then that's a different story.

"EVERY MAN FOR HIMSELF"

I recently helped put on a drill for multiple stations to participate in. We had an acquired structure and pumped smoke into it from a burn barrel on the outside of the building.

The room where the smoke was pumped in had traffic cones spaced out to give the nozzle man a target to flow water at.

Crews were met with moderate smoky conditions upon entry along with some pieces of wood at waist height, which encouraged the advancing team to stay low for visibility purposes.

Crews varied with performance. During one evolution, the nozzle man entered the house performing da' clamp slide (the nozzle forward technique). The backup firefighter carrying the irons walked in behind the nozzle man not assisting with the line. The nozzle man made one turn after entering the front door and was struggling with the advancement. The backup man was still behind him and not assisting. The officer was at the front door, not assisting and talking on the radio.

This crew was not performing as a team because they did not train as a team. They were all focused on their individual tasks and not their function as a fire attack team.

This drill helped make them aware, and they performed as a team on the next evolution.

Many hands make light work.
John Heywood

Practice as a team and you will perform like a team. We must all know what the other person will be doing before the bell rings.

Kitchen table stories

The firehouse kitchen table is where all the departments and even world problems are solved. Many things are discussed at this table. We will only dive into the learning opportunities that come up for now. We learn best from stories. Stories help us paint a mental picture of the event and assist with retention.

Past experiences are good teachers

This is a great opportunity for the senior man to share the good and the bad they have learned on the fire ground over the years. They say we learn from our mistakes, but it's best to learn from others. If we do not learn from our past, we are doomed to repeat it.

Not only can old war stories be entertaining to listen to, but a lot can be learned from them. Whatever you learned on the fire ground can be learned by someone listening at the kitchen table. The more we share our experiences, the more value we provide to others. Give as much knowledge away as possible.

The value of this sharing came to light recently when a member of my crew shared that one of my past experiences at an attic fire gave him insight and knowledge on what to do at one he had recently at a different station. You never know what you may share one day that may help out another firefighter or even save their life.

Therapeutic chats

Therapy? There's nothing wrong with taking care of your mental state. More and more now we are seeing that PTSD is appearing in firefighters. Years and years of exposure to stressful incidents builds up. It is important for us to recognize this and do preventive maintenance on ourselves.

One method of this is to talk about your experiences with others. It's best to share with the people you were on the call with as soon as you get back to the firehouse. Mental health is a serious issue that isn't getting any better unless we do something about it.

Some departments may use a CISD team to come in and go over the call with the group who responded.

Let's face it, though. Oftentimes guys refuse to share or clam up in this environment. If nothing else, share amongst your crew in your station. What might not affect one guy will affect another.

Be there for your brother or sister to listen.

"THE CALLS THAT STICK AROUND"

I had a SIDS call early in my career. This call went to shit. The mother was hysterical and would not put the child down. A chief officer who showed up ended breaking down and left to go be with his grandchild.

The next day my crew ended up at a bar self-medicating.

Next thing you know we are in the parking lot exchanging unpleasantries with some steroided out softball players who had compensation issues. Thank God for a nearby police officer who decided since we weren't going to play nice that everyone needed to go home.

The same year, I ran a triple fatal car wreck that involved individuals my age. My partner and I were assigned to remove the bodies and place them into body bags. I didn't sleep for two days. People didn't talk about calls or see the value in it back then.

Thank God for having a dad in the fire service. He was the one who talked with me and helped me get my mind right.

This shit builds up. Don't let it.

Be there for each other in your firehouse. Each call affects people differently. The older I get the more certain calls get to me.

Tim Kennedy recently shared his perspective about PTSD recovery

You get up early and train. You train so hard your hands bleed, and you sweat acid. You train so hard you collapse seeing stars. You go get cleaned up. Have a healthy meal. Look your best, dress nice. Then know that the real work is about to start. Find something bigger than yourself and pour every ounce of who you are into it. If that's your family, be the best father on Earth. If you are a cop, firefighter, or a trash man, be the BEST.

The full post can be viewed here:
https://www.facebook.com/TimKennedyMMA/posts/1078650138 838938:0

Active people are healthy and sedentary people are not. We are just like our boats and vehicles in that if we do not use them they break down. Exercise your mind and body as often as possible.

Use a stopwatch

A stopwatch is the instructor's secret weapon. I learned about using a stopwatch during a drill from an officer out of the Miami area.

It yields measurable results

Time is black and white. Time does not care about your intentions or feelings. Time only cares about results. Fire doesn't care that you're a good person and tried hard to put it out. A fire is going to grow as fast and as big as it can. We must beat it with time and gallons per minute (GPM). The faster we are ready to fight, the smaller the fire will be.

The missing child's parents don't care that you'll eventually find their child. They want you in there yesterday, bringing the baby to safety in the fresh air.

Timing a drill gives the firefighter feedback on their performance.

A drill is different than a training. We train to build skills, we drill to measure proficiency.

Breed healthy competition

There's a lot more stress on the fire ground compared to when we perform drills in our firehouse. How can we recreate the same pressure? Record the time and add audible distractions. Time and audible distractions should only be included in your drill after the person shows proficiency. What happens when you add more stress to someone who is already stressed? Meltdown, aggression, and frustration. Be sure to ensure proficiency in the skill before you add stressors.

No one wants to let the guy or gal next to them down. We all talk a big game, but you can't talk your way out of measurable results.

How long does it take for you to mask up? Ask your crew your next shift. Sounds like a simple skill, right? There's a big difference, though, between 20 seconds and 60 seconds.

I did this drill with my crew awhile back. Our goal was for all of us to be masked up in 25 seconds. When we started, some times were 50 seconds plus while others were already sub 25 seconds. By the end, everyone was masked up within 25 seconds. Repetition is the key.

Encourage self-reliance and self-belief

We are less likely to want to do something that we are not good at. So let's get good at them. I watched the confidence level of my crew soar after the mask up drill. Some of them cut over 30 seconds off of their time. All it took was enough repetitions for them to figure out a system that worked for them. Some used the Cali flip method and some took a knee. They found a system that worked for them to get the same results.

Once people have their system, they can continue to be successful repeating it. We all like to perform well. Set your people up for success by encouraging them to have a system for doing certain tasks. Mask up the same way every time, put your gear on the same way, check out your SCBA the same way, and grab the same tools.

A quick example of this for me is the TIC. On my engine the TIC is located by the officer. Some leave the TIC in the charger and grab it as needed. I prefer to clip it to my SCBA. This way the TIC is always with me if I have my SCBA on. Chances are high that I'll also need the TIC when I'm using the SCBA. I have now reduced my margin of error of forgetting the TIC if the call turns out to be a fire.

It's not the student; it's the teacher

That guy just doesn't get it. He wasn't listening to me. Good luck with him. Have you ever heard these phrases used to describe a recruit or probie? Maybe at times this may be true.

I have found the majority of the time that the issue is not with the student but with the teacher. Communication issues can be found in any situation that involves conflict.

Many struggle at sharing information when we are good at something and the other person is not. We tend to teach or talk to them like they have the same knowledge we do. Well, they don't. That's why they are the student and we are the teacher.

We must learn to break down the information in a way the other person can understand it. This may involve breaking a complex task down into multiple simple tasks or using visual aids. It's your job as the teacher to make sure people understand the knowledge you are sharing or skill you are teaching.

If people can't understand, it's your explanation

If you regularly find that people are not meeting your expectation, not following directions, or not performing how you ask them to, there's a good chance they did not understand your message.

Sure, there are times when it may be the individual's fault, but most of the time it is the teacher's.

"DO AS I DO, NOT AS I SAY"

I saw this first hand when I had a new guy assigned to me. Firehouse talk about this guy was that he was lazy and struggled with a hydrant. No one seemed to be doing any type of other work with him except for hydrants. The stories I was hearing just weren't adding up to me. Guys screaming at him to apply extra pressure and the lazy comments were coming from lazy guys themselves.

The guy started on my crew and we threw him into the mix. Right off the bat during our expectation talk, I discovered that he was a visual learner. Eureka! A big hurdle was overcome. I am also a visual learner. I learn best when a task is demonstrated to me, not just explained. I struggle with long training outlines or emails with paragraphs of directions.

Try having a student explain what they do know, and go from there.

I got his side of the story, and he was shocked to hear what was being said about him. No one had been clear on his deficiencies to him. Shocker. You mean someone had an issue with someone in the fire service and did not go to them first? That never happens (sarcasm).

I had him show me his skill level at tasks varying from pulling a line, gearing up, masking up, throwing ladders, using the irons, etc. Now we had a base line to start from. I would then demonstrate the system I used to perform a skill. Sometimes I would give multiple ways and let him pick which was best for him. Guess what? His skills started improving. I didn't talk him through the skill, I didn't tell him how to do it. I showed him.

This same individual is now referred to as "squared away". In a short period of time he has become a mentor to more junior firefighters. It was the teaching style that was being used on him that was the issue, not the student.

When teaching, demonstrate the skill you want performed.

Use a whiteboard

Whiteboards or chalkboards are a clear way to get your message across to a group.

PURPOSE	REASON
Focus on board and listen	When giving a training or explaining a skill, the whiteboard literally paints the picture for the listener. This allows the listener to listen to what you are saying and see what you are explaining. The more senses that are stimulated in a training, the higher the retention rate.
Visual representation	Let's say you are explaining what to do at a fire on the third floor of an apartment. The fact you are communicating the plan ahead of time is great. Incorporate a drawing of the "play" and now you're moving in the right direction. If you want to make this "play" concrete, find this building and practices the "play".
Collaborative group thinking	The white board also helps you to come up with multiple ideas. We recently had an outside

	company come in to our firehouse and poll us on what we would like to see different. We did not come up with many ideas as we sat around the table having this discussion. Once I put up some categories on the whiteboard and started writing down what people said, we ended up filling the board. Our brains are meant for coming up with ideas, not storing them. Get your thoughts out of your head and on paper if you want to see anything happen.
Helps ensure everyone's views are heard	Writing down everyone's ideas and thoughts on a subject also helps see everyone's view point. People shy away from sharing their thoughts in fear of being viewed as dumb. Start with yourself and go around getting everyone's input. Odds are that the crew they were last with did not appreciate or value group input.

Videos paint the best picture

Record your training events. A video is an amazing tool to learn from and gain feedback. If you come from a sports background, then you have probably already been exposed to this. Watching footage of how others or you perform helps you become better.

Record your crew

Record your crew during trainings or drills so they can see what you saw. They might see an area that needs improving that you did not. These videos can also be used later on to demonstrate skills to others. Learn and share as much as possible from each other.

Learn from YouTube

You can learn anything from YouTube. How many times have you gone there for a home project idea? More and more videos are being posted from structure fires and trainings held all over the world. What do you want to train on this shift? Forcible entry? Hose line management? Search YouTube for a video on it, watch it as a group, then go out to the bay floor and practice. This cuts down on a lot of prep work that the instructor would normally have to do.

Watching someone else perform at a fire can give great insight on what to do or not to do if you have a similar fire. Understand that unless you were there, you are not getting the whole picture. A lot can be learned from these videos, but being a Monday morning quarterback is a bad habit to get into.

Show them rather than talk about it

Talk is cheap. Show me. Discussion has its place in the training world, but understand that more demonstrations of skills should be going on than just talking about them.

Online versus hands-on

People learn by doing. The online world has enabled us to share everything with just a click of a button. This has proven to be a great tool with sharing information. However, remember, information is shared in the online world, not mastery. That is still done on the training ground with a healthy amount of sweat.

Watching is helpful

There are many resources out there to gather information from. Many departments utilize software to deliver training and have it recorded easily. This can be a double edged sword if you let it. If the information learned from online training is not put into practice, then learning did not take place, only listening.

Firefighters sitting at a computer do not get better at pulling lines, forcing doors, throwing ladders, or being leaders unless they are doing those skills. Too often these online training programs are used as a crutch to think skills are being developed, a checkbox of completion that the person has 20 hours of training done.

Look for results not completions. Unfortunately, there are fire stations around this country where a company officer only leaves their air conditioned office to tell their crew to complete their online training. This officer views this as training and can check it off of their checklist. To me, this act is like wiping before you poop. It makes no sense.

Practice makes perfect

The fire ground is hot, smoky, and stressful. You can add dangerous to this list if the only way you are preparing is in an air conditioned room on a computer. The more prepared you are for something, the less risk there is. Risk does not completely go away, but you can reduce it by being skilled.

Do you watch any of the MMA fights on TV? Those athletes train their asses off and fight. They are prepared for the match. Now, imagine yourself being put in this ring. Do you have a higher risk of injury as opposed to a trained fighter who has been preparing daily? Absolutely. I would be knocked out within the first 10 seconds. A trained fighter mixes up defensive and offensive moves to survive in the octagon. There is still risk for the fighters but not as much risk compared to someone who doesn't prepare themselves. This mindset of preparation should be adopted by all of us.

Proper preparation prevents piss poor performance

We say *everyone goes home* and to be safe, but what are you doing to accomplish this goal? The more we prepare ourselves, the safer we will be.

Let's make this simple: Those who care about their crew's safety will prepare them with training.

Training is typically seen as a way to increase performance, a way to become better. What is an excuse we hear to avoid training? *Well, the fire went out. No one got hurt. My crew is good enough.*

Let's change our view from performance to safety. The better prepared we are, the safer we are. Fewer mistakes happen when we are prepared. Fewer injuries happen when we are prepared. People's lives get saved when we are prepared.

Where do we start then? The basics.

Train on the basics

Everything needs a good foundation to be able to build anything on top of it. The same goes for firefighters. If a firefighter takes two minutes in the front lawn to mask up, then their foundation is shit.

Get a solid base

Start with common tasks performed by a first arriving engine company at a fire

- gear up, seatbelt, airpack
- pulling hose line to door
- masking up
- forcible entry
- hose line management (communication, positions, chasing kinks, pinch points)
- fire behavior
- overhaul/salvage
- investigation

There's more you can come up with, but this is how I broke it down to provide an example.

There are eight listed tasks. Depending on skill level those tasks may take eight shifts or eight months to become proficient. Break up complex skills like first due function into smaller individual tasks. Repeat these tasks until proficiency is demonstrated. Once proficient, you can add a time element to add some pressure. This is an important element since the training ground is more relaxed than the fire ground. Start putting the individual tasks back together one by one.

This concept can be repeated for everything we do. Just like video games, once you pass level one, you move onto level two. This repeats until you beat the game.

You don't know what you don't know until you know what you don't know

Recognize gaps in understanding. Not everyone has the level of training or experience you do. Some may have more and some may have less. Keep this in mind when teaching or drilling.

Here's an example of learning the hard way.

"YOU'VE BEEN DOING IT WRONG"

After a 20-hour class of hose line management taught by passionate instructors, a company officer went up to one of their top chiefs and told him that we had "been doing it all wrong". This chief officer heard, "You've been doing it wrong for over 20 years".

What do you think his reaction was?

If it took 20 hours of hands-on training for you to learn a new skill and gain a better understanding of something, then others will not understand this skill set until they receive the same training.

The same goes for knowledge. If you sat through an eight-hour, smoke-reading class, then why are you expecting someone who wasn't there to have the same knowledge?

We have to share our skills and knowledge with others. This can be difficult since the fire service is full of a bunch of alpha males who all know how to do their job. They do know how to do their job to the level of knowledge and training they have. Increase their knowledge and training, and you'll increase the level they are able to do their job.

Point out deficiencies

You will see deficiencies at trainings, drills, and on the fire ground. If the person is not made aware of the issue, then they will repeat the act.

Provide feedback

There's an art to giving corrective advise to people, and everyone seems to be different. The better relationship you have with a person, though, the easier the feedback will be taken. After action reviews after a drill or incident can sometimes help if the people involved are already aware of the area needing improvement. You will need to provide input, though, if the mistake is not brought up.

"LOW PRESSURE"

We recently had a drill that involved two engine companies responding to an acquired structure for a report of a fire with a missing child. The first due nozzle man and back up man had a hose line that was kinking and folding on itself.

During the after action review, I asked the nozzle man if he had checked his pressure. He said he had and informed his back up man of the low pressure.

The backup man removed some kinks and saw an increase in pressure. However, the nozzle man did not recheck his flow before entry and so they had difficulties.

The engineer was asked what pressure they had pumped the line at.

He responded that the correct pressure was 111psi but he had it at 80psi because he did not want to beat up the guys and it was just a training.

Boom!

Does this engineer know how to do his job? Yes, he does.

Does he lack an understanding of the importance of hose line pressures? Yes.

This engineer believed he was doing the right thing. However, he did not communicate his change in performance to the officer or firefighters. An attempt was made to communicate the importance of proper hose line pressures. Lack of pressure creates kinks and a low GPM which actually makes it harder for the hose team.

Unfortunately, the engineer felt attacked and got defensive.

This will happen. Would I address the issue again? Yes. Would I try to find a better way to share the knowledge? Yes.

Do not be overly critical, support getting it right

The goal of an after action review or critiquing an incident is not to find faults but to encourage proper performance. There is always room for improvement at everything we do. Focus on the act, not the individual.

Someone with an over-inflated ego will most likely take offense no matter how you go about it. We have forms for that.

Google: Butt hurt report.

Only imagine one thing at a time

Alright, guys, we are on scene of a three-story apartment complex with fire on the second floor and a victim hanging from a third story window. Go! Meanwhile, you are sitting in the back parking lot of a strip mall imagining this extravagant scenario that was just explained to you.

Make everything real apart from one thing

The less the student has to imagine, the more they will get out of the training.

"REAL AS POSSIBLE"

Here's an example of a recent drill I was a part of. We obtained an acquired structure that we were unable to burn in. A burn barrel was utilized on the outside of a window, and smoke was pumped in via a dryer vent.

The room that was being filled with smoke had traffic cones in it on the floor, hanging from the ceiling, and just outside the room.

Participants were told the cones were fire. A knocked down cone meant it was out. Crews advanced hose lines under smoky conditions and flowed water at targets at varying heights and in different areas.

Demonstrate training skills first

Whatever you want them to do, show them first. The best way to deliver a new skill set is to demonstrate it first.

Demonstrate clearly

Talk about the skill as you demonstrate it properly. This reduces misinterpretation. You can also have a participant demonstrate it after you and talk about the process as they do it. Here's a tip: start with the most junior member in order to give senior members an opportunity to have a better understanding so they do not get butt hurt if their skills are lacking.

Handle issues as they happen

People may struggle with the skill. When they do, pull them to the side and break down the skill in an effort to locate what area needs correcting. At times it may be as simple as the way they are standing or supporting their weight.

Have them show you as you help

Have the participant demonstrate the skill with you. Help until they show proficiency.

Ask them to show you

Once the skill is understood, have the participant demonstrate the skill with no input. You can even put together a drill with a scenario to test their level of understanding and application.

These strategies will help you make the most of your current workforce.

When people do not want to train it's because they don't want to show how little they know.

No one likes to be embarrassed. No one wants to suck.

Keep people from sucking by training with them so they will not get embarrassed when they have to show what they know instead of just talking about it.

Summary
The secret to increased morale

1. Provide realistic, relevant hands-on training.

2. Train as a team to perform as a team.

3. Share your experiences.

4. Not everyone gets the same amount of jobs or have performed fire grounds skills as much as someone else. On the job training is great, but not everyone receives. Don't wait for experience; go get it with training.

5. A stopwatch is a great tool for drills.

6. People learn differently.

7. Adjust your teaching to the learner.

8. Use visual aids to help with understanding.

9. Start with the basics.

10. Master them first.

11. You don't know what you don't know until you know what you don't know.

12. We are all ignorant to something until we are educated on it.

13. Demonstrate skills first to set a benchmark of what to work towards.

Strategy 7
Leave the fire service better than you found it

Our time here at the fire department is leased. It has an expiration date. The fire department was here long before us and will be here long after us. You are either improving it or weakening it. There is no in-between.

Think how the quote below relates to your department.

> *Out of every one hundred men, ten shouldn't even be there, eighty are just targets, nine are the real fighters, and we are lucky to have them, for they make the battle. Ah, but the one, one is a warrior, and he will bring the others back.*
> *Heraclitus*

Where do you fit in this mix? Will you be the one to bring them back?

You can start by training your replacements to be better than you. In this chapter I will show you how by

- setting the right mindset
- building a teamwork culture
- sharing the experiences you have had

Coach

Coach new recruits, probies, new guys, experienced guys, and veterans. Every team has a coach. They are there to help the team achieve victory. The coach is not the one throwing the ball, making the touchdowns, or grabbing interceptions. That's the players, your crew. A coach can ruin a team's season or make the underdogs dominate.

Motivate

Keep your crew's spirits high. You accomplish this by having high spirits yourself. Motivated individuals perform at high levels. They got there with hard work. Train your crew to go to high levels and they will stay motivated.

Stay on the right path

Help your crew with setting goals and staying on the right path to achieve them. Too often people are left on their own to become better. That's not leadership, it's laziness. You don't have to do the work for them, but you should be showing the way and reminding them where they are at in the process.

Stay open to feedback

Show you are listening by considering input from the crew. The crew may be seeing something you are not. They will look out for you if they know you are looking out for them.

Develop

Develop other people by giving them competence and clarity.

> *"If you come to me with a problem, then you better have two solutions."*

Ever hear this before?

How much of your day is spend solving other people's problems? Imagine being surrounded by people who come to you with the solutions to their problems and are just telling you so you are aware?

This can be done, but everyone needs to be trained to the level of competency needed and have clear directions.

Give responsibility

The key to development is to give up power by giving more responsibility. The more responsibility a person has, the more value they feel they have to offer. Every department's rank structure is different; I will share how I do mine around my firehouse.

My engineers are in charge of the apparatuses and everything on them. If something is broke or missing, they are to fix it or have it replaced. I should be kept in the loop, but they handle it. I also view this rank as an important part of our leadership hierarchy. I expect the engineers to be handling issues at their level and only coming to me if they are unsuccessful with handling the issue.

If I notice some discrepancies, I go to my engineers to handle it first. This is successful since they saw me handling issues first before I started going to them. You develop people by building them. Start off with smaller responsibilities and build on from there.

You will be surprised how much people grow when they are given more responsibility. Prepare them for the responsibility you are giving them. They may need assistance with it from time to time. No one hits the ground running; we start by crawling.

Solve your own problems

Most problems should be solved at the lowest level possible. When every issue goes straight to the officer, this is like little kids tattling to Mom and Dad when lil Joey won't share his toy.

You can enable them to solve their own issues by helping them figure out what their goal is and let them implement it.

I wrote another book another book (available for free at www.brotherhoodcoaching.com) just on this topic and how I implemented it in my firehouse with success.

The original concept is from the book Turn the Ship Around by David Marquet. In this book, David implemented the ladder of leadership.

THE LADDER OF LEADERSHIP

	WORKER	LEADER
7.	I've been doing...	7. What have you been doing?
6.	I've done...	6. What have you done?
5.	I intend to...	5. What do you intend...
4.	I would like to...	4 What would you like...
3.	I recommend...	3. What do you recommend?
2.	I think...	2. What do you think?
1.	Tell me what	1. Do this...

You can also go to the following two sites and watch a video which explains the method

- https://www.youtube.com/watch?v=DLRH5J_93LQ
- https://www.youtube.com/watch?v=-sri5wyth4I

Everyone has strengths

Recognize the strengths that each member of your crew has. We all grew up different and have had different experiences. Some may have a construction background. Here's your go to guy around pre-plan time, building props, or putting on training for others. You may have someone who is mechanically inclined. You now have a "Mr. Fix It" on your crew.

How about a guy who is always smiling and happy? Here's your mascot. This will be the guy who reminds you about the good things that are going on and not to dwell on the bad.

Maybe you've got a guy extremely into health and fitness. Here's the crew's personal trainer.

Good cook makes a firehouse chef

The examples can go on. Now, all these examples of people may also have some weaknesses. We all have them. The important part is to remember to use people in the areas they are strong in. If you force someone to become better in an area they are not skilled in, they will get frustrated and give up. If you always judge them by their inability to cook a good meal, then they will think they suck overall even though they excel in other areas.

We are a team for a reason. We are stronger together.

Raise them to be successful

Teach them everything you know. Why learn the hard way when we can learn from others?

Share what worked

Share what has worked for you and what has not worked. We will never get anywhere as a profession if everyone makes the same mistakes as their predecessors.

Fail safely

Never set your people up for failure. Give them the best guidance you can. There will be times when they must learn for themselves though. This is best done on the training ground.

Let's say someone wants to force a door a way you know will fail. Let them. They can now see for themselves. We have to know when to let people learn for themselves and when to keep them from making mistakes.

Hard lessons are learned on the fire ground. This can be prevented by spending more time on the training ground.

Say failure is ok

Everyone fails. It's part of life. No one is perfect at everything. Most people give up at something after their first failure: promotion tests, training, marriage, their passion. When we give up, that's it; game over. Teach them that success comes from failures. The only people who truly fail are the ones who give up and quit.

Give resources to learn

Expose your people to as much knowledge as possible. Now, there is a difference between knowledge and information. Information is generic busy work that most crews are given to review in order to suffice a check mark in a box. Knowledge is impactful, valuable, and makes a person better after learning it. Knowledge comes from our experienced-based predecessors.

Think of it like this: do you want to learn from someone who has been there or from someone who just repeats a curriculum they were given so everyone gets a certification?

Our career is full of certification classes. There are fire instructor I-III, fire officer I-IV, safety officer, public information officer, and the list goes on. What did you learn from them? I never had any of those *aha* moments in them. You know, when you learn something that just makes everything click.

Maybe it's just me, but I have learned more attending classes from greats like Dave McGrail, Bill Gustin, Curt Isakson, Aaron Fields, John Salka, and others who just share what they are doing.

Encourage attendance to classes taught by the people making an impact in the fire service. Can't make their class, watch their webinars, read their books, or read their articles? Some of these guys will give out their information and put out open invitations so you can go ride along on their rig with them.

Adversity builds a team

Ever notice the bond found in a family? A family is strong because of all of the challenges they have overcome together. Families experience deaths, injuries, illness, layoffs, heart aches, and when they overcome these obstacles of life, they become stronger. This is seen in the military when a platoon sees combat together. They are fighting for their life and the life of the person next to them. How can they not become kin after this?

If you have not seen this yet, you will notice as firehouse crews experience more and more stressful calls together, they become closer. That bond between them becomes stronger. Everyone has proven themselves on the fire ground that they are there for one another. We can start building this bond before the bell rings though. The challenges we face while training will make you work closely together and develop stronger and tighter relationships.

All the examples above start with a challenge which is overcome by work performed by a team of individuals which then increase their bond as a team.

Stressful Challenge + Work + Team =
Team Building (Family bond)

Conversations get easier

You need to tell people when something's wrong. Let's face it, the norm is to not upset others. We beat around the bush and speak so indirectly to not hurt someone's feelings that we end up making the situation worse. When we speak to someone indirectly, odds are that they will not understand what we are trying to communicate.

You can't stay silent

Many choose to stay silent over issues when they arise. They bury their head in the sand like an ostrich. What kind of message are you sending to the hard workers when you let the lazy ones slide? The hard workers notice and they remember.

The more you have these uneasy conversations, the easier they become. It sucks in the beginning, but you end up hurting the ones doing all the work or doing the right thing when you don't have the conversation with those who are not.

A practice that makes it easier is to role play with the person. Show how their action or inaction hurt or gave a crew member more work. Nobody wants to let their crew members down. Most do not see how their decision impacted others. Just by making them aware can impact future decisions this person will make.

SAMPLE SCENARIOS

Someone is late to work. Someone from the shift before has to stay over.

Not wearing proper PPE on a call. Everyone else has to work harder since they won't be ready to work. Risk of injuring self.

Not knowing proper hydraulic calculations. Fire attack team has less water, kinks in hose, and difficulty maneuvering hose line. Risk of injury to crew increases.

These are just a few. There a whole bunch more. The one thing that crews have in common is that no one wants to injure the guy or gal next to them. Show them how the decision they made hurts the team to make an impact in their future decisions.

Positive reinforcement

It is crucial to recognize when a good job is performed. A person is more likely to repeat an act when they receive praise for it. There are a lot of books and studies about positive reinforcement. I stumbled across it in person one day and saw the benefits.

We have computers on our engines which show dispatch information and a map to the call. This is new within the last five years or so. Before then, the engineer would reference a map book when they weren't sure where the address was. I cannot remember the last time I saw an engineer reference the map book since we got the computers.

A call came in to an address that wasn't familiar to us. When I hopped into the engine, I noticed my engineer had the map book out, looking for the address. He didn't wait for me to read the map on the computer. He took the initiative to know where we were going beforehand. After the call, I mentioned to him that it is a rarity to see the map book out and I was impressed with the initiative he took to figure out the location himself as opposed to waiting for me to look it up. His response was that it was no big deal and he was just doing his job. He is a modest person and doesn't like attention. Well, lo and behold, that map book was out and being reference by him on the next two calls.

That's when it dawned on me. Who wouldn't like being told they are doing good or excelling above others? Be sure to be specific with positive reinforcement. A phrase like "good job" is not specific enough. Try "I like the way you used the map book to find the location of the call sooner" or "I appreciate you taking the new guy under your wing and showing him some forcible entry techniques". Be specific so the act of positive reinforcement is impactful.

Let the boys handle it

A leader who leads by example will eventually be able to step back and watch their crew start stepping up. Don't keep stepping in first when problems arise. This is great to start around the station.

Let them be responsible for method

As long as the end goal is the same, do not worry about how they accomplish it. We all have a different process for how we get things done. Understand that yours may not be the best way for everyone. Let your crew use a method that's best for them.

Give them a chance to learn

When we step back and offer guidance only when asked, the learning process increases.

A common example in most firehouses are issues with the probie.

Your crew will determine if the issue goes to the officer or the crew handles it, depending on their culture. People like to report things to the officer right away instead of handling it themselves or checking with the "senior man". I started to hear some "rumblings" about one of the probies who was assigned to me. I was not too worried since no one on my crew had come to me with any issues. I did check in with one of my senior guys to ask if there were any issues.

He shared with me some items that they did not like about the probie. I asked him if it was at my level yet or if they were handling it. He said that they were handling it and would let me know if they needed me.

This is the point when I realized we were doing better than I thought as a crew. I knew we were a solid crew by the results I was seeing, but it was this interaction that made me realize how lucky an officer I was to have a crew like this.

Discourage escalation for the sake of it

There are a couple benefits when crews handle issues before they start using the chain of command. Anytime an officer has to be involved with performance issues the whole interaction is more stressful for the person being talked to.

They may view the conversation as an attack or be concerned that paperwork will soon follow. When the crew handles the situation it is kinda like a big brother looking out for you. The crew gets a chance to mentor the person, and the person feels accepted by the group since they are looking out for them.

I have found that most issues are best resolved with conversation and getting buy-in by communicating a better understanding of the issue. There will be a time, though, when ego comes into play and someone just downright refuses to play well.

Help them improve deficiencies

If someone is deficient in a skill or knowledge, then help them. What do you do when you see an elderly person on the ground after falling? Do you tell them they should be more careful and learn how to walk better or do you help them up? Anyone can point out a person's flaws. It's called being a critic.

Recognize deficiencies

When there's a deficiency with someone on your crew, help them recognize the issue. Work together to come up with a plan of improvement. This could be some hands-on training or some further knowledge of the issue. If you noticed the deficiency, then you are aware of a better way, so share what you know. Too often people are told to seek their own help. There has to be ownership with both parties involved.

It's up to the person to do the work to become better, but the leader needs to help set them up for success. This can be done with a plan with benchmarks, setting up a drill to test what they learned, or actually teaching the skills needed.

When people are sent off on their own, we leave their success up to chance. When we set them on the right path, we guarantee success.

Push forward, don't hold back

Encourage progression of your people.

Give positive feedback

If someone is not performing at the level you want them to be at, you can either accept the level they are at or help them improve. No one gets better by holding them back. This is commonly seen when someone is working to achieve the rank above the one they have. Many hold people to a higher standard than they were at when it comes time for releasing someone to work at a higher rank. Phrases like "You are not quite there" or "Let's have another officer evaluate you" are ways of holding someone back.

Sure, some may figure it out themselves, but you can speed things along if you actually work with the person.

People need regular feedback to know how they are doing when taking on new skills. They can make the corrections needed when they are aware what needs to be done.

Summary
Leave the fire service better than you found it

1. Coach those around you – motivate.

2. Develop those around you – give responsibility.

3. Be present with your crew.

4. Focus on each person's strengths.

5. Set those around you up for success.

6. Adversity builds a team.

7. Communicate when something is wrong – don't bottle it up to blow up later.

8. Provide positive reinforce often.

9. Let your crew handle issues they are able to.

10. Push people forward, don't hold them back.

These are some ways to develop the fire service of the future. We all learn from those who came before us.

For some these were good examples and for others these were poor examples.

The only way to make things better is to work hard on providing the best example you can for others to emulate.

Bibliography

Book

Keller, Gary, and Jay Papasan. *The One Thing: The Surprisingly Simple Truth behind Extraordinary Results*. Austin, TX: Bard, 2012. Print

Marquet, L. David. *Turn the Ship Around!: A True Story of Turning Followers into Leaders*. New York: Portfolio, 2012. Print.

Shaw, Dan, and Douglas J. Mitchell. *25 to Survive: Reducing Residential Injury and LODD*. Tulsa, OK: PennWell Corporation, 2013. Print.

Viscuso, Frank. *Step up and Lead. Tulsa, OK:* PennWell Corporation, 2013. Print.

Willink, John, and Leif Babin. *Extreme Ownership: How U.S. Navy SEALs Lead and Win*. New York: St. Martin's, 2015. Print.

People

Fields, Aaron. Firefighter. 2016

Vonappen, Mark. Firefighter. 2016

Castellanos, Oliver. Firefighter. 2016

Dusick, Brad. Firefighter. 2016

Films

The Ladder of Leadership. Dir. Mile Madina. Perf. David Marquet. *YouTube*. YouTube, 17 Sept. 2014. Web. 15 Dec. 2016.

Marquet, David. "How Great Leaders Serve Others: David Marquet at TEDxScottAFB." YouTube. YouTube, 21 June 2012. Web. 15 Dec. 2016.

Social media
Facebook

Kennedy, Tim, 2016. Facebook. August 31st. Available from: facebook.com/TimKennedyMMA/posts/1078650138838938:0

Resources

If you want to download any of the resources mentioned in the book, here are the links.

1. Downloadable your templates at:
    http://brotherhoodcoaching.com/fix-your-firehouse/

2. Fighting fires not fighting fires around your firehouse book and class.at:
    http://brotherhoodcoaching.com/lighting-fires-not-fighting-fires-around-your-firehouse

3. 24/7 access to brothers like you in the Brotherhood Coaching Facebook group at:
    https://www.facebook.com/groups/BrotherhoodCoaching/

Keep in touch

I hope you have enjoyed learning about experience-driven strategies for your firehouse.

I wish you every success.

Share your stories with me. I'd love to hear from you.

Email them to

- john@brotherhoodcoaching.com

or join my Facebook group and share them with other firefighters.

- https://www.facebook.com/groups/brotherhoodcoaching/

Be the role model you looked up to,

John Lovato, Jr

www.ingramcontent.com/pod-product-compliance
Lightning Source LLC
Chambersburg PA
CBHW071848200326
41519CB00016B/4291